Thilo Pollak

Influence of side walls and undulated topography on viscous film flow

Thilo Pollak

Influence of side walls and undulated topography on viscous film flow

Südwestdeutscher Verlag für Hochschulschriften

Impressum / Imprint
Bibliografische Information der Deutschen Nationalbibliothek: Die Deutsche Nationalbibliothek verzeichnet diese Publikation in der Deutschen Nationalbibliografie; detaillierte bibliografische Daten sind im Internet über http://dnb.d-nb.de abrufbar.
Alle in diesem Buch genannten Marken und Produktnamen unterliegen warenzeichen-, marken- oder patentrechtlichem Schutz bzw. sind Warenzeichen oder eingetragene Warenzeichen der jeweiligen Inhaber. Die Wiedergabe von Marken, Produktnamen, Gebrauchsnamen, Handelsnamen, Warenbezeichnungen u.s.w. in diesem Werk berechtigt auch ohne besondere Kennzeichnung nicht zu der Annahme, dass solche Namen im Sinne der Warenzeichen- und Markenschutzgesetzgebung als frei zu betrachten wären und daher von jedermann benutzt werden dürften.

Bibliographic information published by the Deutsche Nationalbibliothek: The Deutsche Nationalbibliothek lists this publication in the Deutsche Nationalbibliografie; detailed bibliographic data are available in the Internet at http://dnb.d-nb.de.
Any brand names and product names mentioned in this book are subject to trademark, brand or patent protection and are trademarks or registered trademarks of their respective holders. The use of brand names, product names, common names, trade names, product descriptions etc. even without a particular marking in this work is in no way to be construed to mean that such names may be regarded as unrestricted in respect of trademark and brand protection legislation and could thus be used by anyone.

Coverbild / Cover image: www.ingimage.com

Verlag / Publisher:
Südwestdeutscher Verlag für Hochschulschriften
ist ein Imprint der / is a trademark of
OmniScriptum GmbH & Co. KG
Heinrich-Böcking-Str. 6-8, 66121 Saarbrücken, Deutschland / Germany
Email: info@svh-verlag.de

Herstellung: siehe letzte Seite /
Printed at: see last page
ISBN: 978-3-8381-3925-8

Zugl. / Approved by: Bayreuth, Universität, Diss., 2012

Copyright © 2015 OmniScriptum GmbH & Co. KG
Alle Rechte vorbehalten. / All rights reserved. Saarbrücken 2015

List of Journal publications

- Pollak T. and Köhler W.: *Critical assessment of diffusion coefficients in semidilute to concentrated solutions of polystyrene in toluene*, The Journal of Chemical Physics, **130** (2009), 124905

- Wierschem A., Pollak T., Heining C. and Aksel N.: *Suppression of eddies in films over topography*, Physics of Fluids, **22** (2010), 113603

- Haas A., Pollak T. and Aksel N.: *Side wall effects in thin gravity-driven film flow - steady and draining flow*, Physics of Fluids, **23** (2011), 062107

- Pollak T., Haas A. and Aksel N.: *Side wall effects on the instability of thin gravity-driven films - From long-wave to short-wave instability*, Physics of Fluids, **23** (2011), 094110

- Heining C., Pollak T. and Aksel N.: *Pattern formation and mixing in three-dimensional film flow*, Physics of Fluids, **24** (2012), 042102

- Heining C., Sellier M. and Pollak T.,: *Flow domain identification from free surface velocity in thin inertial films*, Journal of Fluid Mechanics, **submitted**

- Pollak T., Aksel A.: *Experimental evidence of multiple instability branches of gravity-driven films over topography*, **in preparation**

Abstract

While a gravity–driven viscous film flow down an inclined flat plane of infinite extent can be described by an easy analytical solution, flow problems in nature, like glacier movements or the liquid film on the human eye are much more complex. Also to optimize a large number of technical applications, like coating applications or heat exchanger devices, one has to investigate and understand how different influencing factors, like topological features on the substrate or a finite width of the system, influence the flow and its stability isolated from each other.

By introducing a wavy structure to the underlying topography, which could be for example a model for roughness, new effects emerge in the flow, which cannot be observed in flows over a flat incline. Eddies can separate from the main flow at the lee side of the undulation for kinematic reasons, or induced by inertial effects. In biological systems these eddies are dead water areas, which are cut off from nutrient supply, in heat exchanger applications their appearance has a strong impact on the convective heat transport within the liquid. Furthermore, the amplitude of free surface of the liquid can be amplified immensely when the liquid is in resonance with the undulation of the underlying topography. In this work we study experimentally as well as numerically the complex interaction of this resonance phenomenon with the appearing of eddy structures in the valleys of the undulation and show, that one can suppress flow separation selectively even at rather high Reynolds numbers when one exploits the resonance phenomenon specifically.

Another part of this work deals with the question how the presence of side walls and the contact angle of the liquid there influences the free surface shape of the liquid, the velocity field and the globally transported volume flux. While an additional no–slip condition at the wall causes additional friction and leads thus to a lower volume flux, capillary elevation at the side walls can generate a velocity overshoot in the vicinity of the walls, depending on the film thickness and the wetting properties of the liquid, which counteracts the additional friction coming from the walls. An extensive theoretical parameter study, which is supplemented with experimental data, provides criteria for the first onset of the velocity overshoot and gives answer to the question when the counteracting influences on the global volume flux just cancel each other.

An experimental study of the free surface shape of a draining flow shows that this configuration cannot be described by a series of quasi–steady states, even when a dynamic contact angle is taken into consideration, although the flow changes only very slowly in time. Additional time dependent numerical simulations of the draining flow reveal an indentation of the free surface in the vicinity of the side wall, which could promote film rupture in technical thin film applications.

Furthermore, side wall effects play an important role for the physical stability of the flow. Waves develop at the free surface of a gravity–driven flow and grow while they are traveling downstream, when a critical volume flux is exceeded. It is shown by experimental variation of the contact angle, the film thickness and the side wall distance, that the presence of side walls generates different effects which have competing influences on the stability of the flow. Capillary elevation leads to a pretensioning of the free surface, which tends to stabilize the free surface, just as the additional no–slip condition at the wall does. The emerging of a velocity overshoot in the capillary elevation on the other hand leads to a destabilization of the flow. In the system studied here the stabilizing influence of the side walls dominates over the destabilizing influence which is of comparatively short range, which means that this flow configuration is more stable than the corresponding flow of infinite extent. However, the results suggest that the destabilizing influences should dominate over the stabilizing influences in similar flow configurations when the film would become even thinner. While free surface film flows typically form long waves at first, we find for this flow configuration, that the type of instability changes from a long–wave type in the middle of the channel to a short–wave type instability, as it is well known for boundary layer flows, as the side wall distance is reduced.

Zusammenfassung

Während sich eine schwerkraftsgetriebene viskose Strömung, die eine unendlich ausgedehnte und glatte Ebene hinabfließt, durch eine einfache analytische Lösung beschreiben lässt, sind die Strömungsprobleme in der Natur, wie zum Beispiel eine Gletscherbewegung oder ein Flüssigkeitsfilm auf dem menschlichen Auge, weitaus komplizierter. Auch um zahlreiche technische Anwendungen, wie beispielsweise Beschichtungs- oder Wärmetauschprozesse, optimieren zu können, müssen Einflussfaktoren, wie das Vorhandensein einer Struktur auf der Oberfläche des Bodens oder eine endliche Breite des Systems und deren Einflüsse auf das Strömungsfeld und die physikalische Stabilität der Strömung isoliert untersucht und verstanden werden.

Durch das Vorhandensein eines gewellten Untergrundes, der zum Beispiel ein Modell für Rauheit sein könnte, entstehen neue Effekte in der Strömung, die bei einem glatten Untergrund nicht beobachtet werden können. Sowohl aus rein kinematischen Gründen, aber auch durch trägheitsinduzierte Effekte kann die Strömung auf der Windschattenseite der Bodenstruktur vom Boden ablösen, so dass in den Bodenmulden Rezirkulationsgebiete entstehen. In biologischen Systemen stellen diese Regionen Totwassergebiete dar, die nicht mit Nährstoffen versorgt werden, in Wärmetauscheranwendungen hat ihr Auftreten einen starken Einfluss auf den konvektiven Wärmetransport. Neben dem Entstehen einer Strömungsablösung kann durch Auftreten von Resonanz zwischen dem gewellten Boden und der Flüssigkeit die Amplitude der freien Flüssigkeitsoberfläche immens verstärkt werden. In dieser Arbeit wird das komplizierte Zusammenspiel aus Resonanz und dem Entstehen von Rezirkulationsgebieten in den Bodenmulden sowohl numerisch als auch experimentell untersucht und es wird gezeigt, dass man durch geschickte Ausnutzung der Resonanz das Auftreten der Wirbelstrukturen auch bei relativ hohen Reynoldszahlen gezielt unterbinden kann.

Ein weiterer Teil dieser Dissertation befasst sich mit der Frage, wie sich das Vorhandensein von Seitenwänden und der Kontaktwinkel der Flüssigkeit dort auf die Form der freien Oberfläche, das Geschwindigkeitsfeld und den globalen Volumenstrom auswirkt. Während eine zusätzliche Haftbedingung an der Wand zu zusätzlicher Reibung und damit zu einem geringeren Volumenstrom führt, kann in Abhängigkeit von Kontaktwinkel und Filmdicke durch kapillare Anhebung ein Geschwindigkeitsüberschuss in der Nähe der Seitenwand entstehen, der dem Einfluss der Haftbedingung entgegenwirkt. Eine umfangreiche theoretische Parameterstudie, die durch experimentelle Ergebnisse ergänzt wird, liefert Kriterien für das erste Auftreten eines Geschwindigkeitsüberschusses und beantwortet die Frage, wann sich die entgegenwirkenden Einflüsse auf den globalen Volumenstrom gerade gegenseitig aufheben.

Die experimentelle Untersuchung der Form der freien Oberfläche einer Drainageströmung zeigt, dass sich diese Strömung auch durch Einführung eines dynamischen Kontaktwinkels nicht durch eine Folge quasistatischer Zustände beschreiben lässt, obwohl sie sich zeitlich nur sehr langsam verändert. Zusätzliche zeitabhängige numerische Simulationen der Drainageströmung offenbaren eine Vertiefung der freien Oberfläche in der Nähe der Seitenwand, die in technischen Dünnfilmanwendungen einen Abriss des Flüssigkeitsfilms hervorrufen könnte.

Darüber hinaus spielen Seitenwandeffekte auch eine entscheidende Rolle für die physikalische Stabilität der Strömung. Auf der freien Oberfläche einer schwerkraftsgetriebenen Filmströmung bilden sich Wellen aus, die anwachsen, während sie die Ebene hinabfließen, sobald ein kritischer Volumenstrom überschritten wird. Es wird durch experimentelle Variation des Kontaktwinkels, der Filmdicke und des Seitenwandabstandes gezeigt, dass verschiedene Effekte, die durch das Vorhandensein von Seitenwänden auftreten, miteinander konkurrierende Einflüsse auf die Stabilität der Strömung haben. So hat die kapillare Anhebung eine Vorkrümmung der freien Oberfläche zur Folge, welche zusammen mit der zusätzlichen Haftbedingung an der Wand zu einer Stabilisierung der Strömung führt. Das Auftreten eines Geschwindigkeitsüberschusses in der kapillaren Anhebung führt hingegen zu einer Destabilisierung der Strömung. Bei der hier untersuchten Strömung überwiegen die langreichweitigen stabilisierenden Einflüsse den vergleichsweise kurzreichweitigen destabilisierenden Einfluss der Seitenwand, so dass dieses System insgesamt gegenüber einer quer zur Hauptströmungsrichtung unendlich ausgedehnten Strömung durch die Seitenwände stabilisiert wird. Die Ergebnisse legen jedoch nahe, dass für ähnliche Strömungskonfigurationen, die eine noch geringere Filmdicke aufweisen, der Nettoeinfluss der Seitenwand auf die Strömung auch destabilisierend sein könnte. Während sich bei Filmströmungen typischerweise zuerst lange Wellen auf der freien Oberfläche ausbilden, finden wir für diese Strömung, dass sich durch eine Verringerung des Seitenwandabstandes ein Übergang von einer Langwelleninstabilität zu einer Kurzwelleninstabilität, wie man sie typischerweise von Grenzschichtströmungen kennt, vollzieht.

Contents

1	**Introduction**	**10**
2	**Experimental systems and setups**	**16**
	2.1 Liquids	16
	2.2 Flow facilities	17
	2.3 Tracer particles	18
	2.4 Experimental setups	20
	2.4.1 Flow rate	20
	2.4.2 Detection of the free surface shape	21
	2.4.3 Streamline detection	23
	2.4.4 Velocity field measurements	24
	2.4.5 Stability measurements	25
3	**Two–dimensional film flow**	**32**
	3.1 Suppression of eddies	32
	3.1.1 Problem formulation	32
	3.1.2 Experimental and numerical findings	34
	3.1.3 Physical interpretation and discussion	42
	3.1.4 Conclusions	44
4	**Three–dimensional film flow**	**46**
	4.1 Basic flow	46
	4.1.1 Governing equations	46
	4.1.2 Flow type classification	50
	4.1.3 Flow rate study	51
	4.1.4 Velocity field	54
	4.1.5 Free surface shape	57
	4.1.6 Conclusions	60
	4.2 Stability near the side walls	62
	4.2.1 Results	62
	4.2.2 Conclusions	65
5	**Conclusions and outlook**	**68**

Contents

Chapter 1

Introduction

Viscous thin film flow configurations can be found in a large number of environmental systems such as the liquid film on the human eye, the flow on a wetted road or the thin water film emerging under objects which are sliding over ice, but also in much larger systems such as glacier movement[1], avalanches[2], lava flows or debris[3]. Furthermore, it is an interesting flow configuration for many industrial systems like spin- or curtain-coating applications[4, 5, 6], heat exchangers[7, 8], evaporators, condensers or absorption and rectification columns. Especially to optimize industrial processes it is essential to understand the underlying physics in such film flows. Therefore, the number of publications dealing with viscous gravity–driven thin film flow is numerous and still growing rapidly, showing a lively interest into this subject.

It was Nusselt[9] who first presented an exact analytical solution of the steady Navier–Stokes equations for a viscous liquid film flowing down a flat incline of infinite extent. This strongly idealized solution is often not able to describe the physics in real life applications, may it be, because the influence of a finite extent of the flow configuration cannot be neglected, or because the substrate is not perfectly flat. This might be due to spurious imperfections at surfaces, due to a finite roughness, or due to undulations which have been added intentionally to the substrate to increase the surface area, as is often useful in technical applications, for example in heat exchangers. Additionally, the physical stability of the steady Nusselt solution becomes a concern at higher volume fluxes. One finds, that a free surface flow is not steady over all volume fluxes or Reynolds numbers, respectively[10, 11]. When a critical value is exceeded, the free surface of the liquid becomes unstable and waves start to develop from infinitesimal disturbances and travel downstream. When the volume flux is increased even further complex wave structures emerge[12], before a transition to a turbulent flow occurs, which is characterized by stochastic behavior[13].

To get insight into the problem of gravity–driven free surface flows over topographies with undulations of finite amplitude, research on this topic has gained more and more interest over the last years. However, since the investigation of such systems involves several technical difficulties, which are mainly coming from the limited optical accessibility due to the curved liquid boundaries at the substrate and the free surface, the number of experimental publications dealing with flows over undulated topographies is still comparatively low. Decré and Baret[14] investigated the influence of two–dimensional step–up, step–down and trench geometries on the free surface shape of thin water films flowing above it by using phase-stepped interferometry. They found their results to be in good

Chapter 1. Introduction

agreement with the theoretical results for the Green's function of the linearized problem by Hayes et al.[15] who studied the influence of arbitrary small substrate defects on thin liquid films. Also Kalliadasis et al.[16] studied the influence of various topographical features like steps, trenches or mounds on thin viscous liquid films, using lubrication theory. They found, that the film dynamics is governed by three pertinent parameters, the depth, the width and the steepness of the feature. Mazouchi and Homsy[17] presented numerical solutions of the Stokes flow case over step and trench features for different surface tensions. Similar to the results of Aksel [18], who studied the influence of capillarity on a film flowing over an inclined plane with an edge numerically as well as analytically, they also found, that capillary forces cause the free surface to develop a ridge before a downwards edge. Negny et al.[19] studied the influence of a sinusoidally undulation at a vertically aligned wall on heat and mass transfer in laminar films flowing above it. Wierschem et al.[20] presented different theoretical perturbation approaches for the limits of thin film flow over weak undulations, thin film flow over stronger undulated bottom profiles and for thick films flowing over weak sinusoidal undulations and also compared these calculations with their experimental data presented in [20] and [21]. However, since real world problems are never purely two–dimensional more and more authors[22, 23, 24, 25] started to focus their studies on film flow over three–dimensional undulated topographies, which was strongly facilitated by the tremendous increase of computational power during the last years. Very recently the investigation of the so called "inverse problem", where not the free surface of the liquid, but the geometry of the underlying substrate is unknown *a priori*, draw much attention due to its excellent technical applicableness.[26, 27, 28, 29, 30]

Pozrikidis[31] presented an extensive numerical parameter study on two–dimensional free surface Stokes flows along sinusoidally undulated walls at different wave amplitudes, inclination angles, flow rates and surface tensions. He focused especially on flow reversal, which appears when a sufficiently thick film flows over strongly corrugated topographies and established a criterion for the appearance of a flow reversal in the troughs of the undulation. Such eddies were later shown to act as "fluidic roller bearings" for the improvement of material transport in creeping films.[32] Wierschem et al.[33] studied the genesis and growth of eddies at very low Reynolds numbers ($\mathrm{Re} = \mathcal{O}(10^{-5})$) experimentally for various wavy topography geometries and mean inclination angles. They found, that under creeping flow conditions not the Reynolds number is responsible for the creation of eddy structures, but a critical film height, which depends on the waviness of the underlying topography, has to be exceeded. Scholle et al.[34] presented an analytical solution method of the Stokes equations based on complex function theory for arbitrary film thickness and waviness. They showed that also higher order eddies can be created, which rotate into the opposite direction of their neighboring eddy, in very steeply undulated geometries. They carried out detailed parameter studies and found their results to be in good agreement with experimental data. These so called kinematically induced eddies appear under Stokes flow conditions and are symmetric as the whole flow field is, because the Stokes equations are space reversible. The influence of inertia on the eddies created by strong topography undulations has been studied experimentally by Wierschem and Aksel[35] and numerically by Scholle et al.[36]. They found, that adding inertia leads to a growth of the eddies and tends to shift them to the lee side of the undulation. Scholle et al.[36] found that the tilting of the eddy in the troughs of the undulation is governed by a local Reynolds number which is associated with the characteristic corrugation length scale of the substrate and quantifies the near–field geometric and inertial influences on the flow structure in the troughs of the undulation. Trifonov[37] carried out numerical analysis of liquids falling on a vertically aligned undulated wall over a wide range of Reynolds numbers and geometries in the integral boundary–layer framework and compared his results with experiments of Zhao and Cerro[38]. At very high Reynolds numbers (Re = $\mathcal{O}(100)$) he found a region with a non–monotonous growth behavior of the eddy size with increasing Reynolds number.

Bontozoglou and Papapolymeru[39] carried out a linear analysis for small–amplitude disturbances to study the interplay between the liquids free surface and a sinusoidal undulation at the wall and found a resonance phenomenon which leads to an amplification of the free surface amplitude. Bontozoglou[40] studied numerically the interplay between such a free surface amplification and the streamline pattern. He found that flow separation in the troughs of the undulation can systematically be delayed at Reynolds numbers of about 200 when the free surface corrugation is roughly in phase with the underlying geometry, as is the case under resonance conditions. Wierschem and Aksel[41] reported about experimentally observed standing waves in films flowing over substrates of moderate waviness when the free surface of the liquid is in resonance with the bottom topography. Furthermore, they found that the undulation of the topography may cause the liquid to form hydraulic jumps and complex three–dimensional surface rollers. Wierschem *et al.*[42] and Heining *et al.*[43] studied linear and nonlinear resonance analytically and numerically and revealed the relevant physical mechanisms, which are similar to resonance known from classical mechanics. The interaction between the free surface of a liquid and eddies appearing in the troughs of undulation was studied numerically as well as experimentally by Wierschem *et al.*[44]. They report on a strong indentation at the free surface of the liquid, which can be identified as a hydraulic jump, when the Froude number of the flow is of the order of one. Furthermore, they show that eddies can systematically be suppressed at moderate Reynolds numbers under steady or weakly unsteady flow conditions, when the amplitude of the free surface is amplified by resonance. Nguyen and Bontozoglou[45] studied flow separation at steeply corrugated walls numerically and compared their results with independent experiments[33, 41]. They found a critical Reynolds number, which divides the flow into a regime with subcritical and regime with supercritical flow separation. For very steep corrugations they found both solution branches to coexist.

As the Kapitza family showed with their pioneering experiments[10, 11], not only a presence of a finite wall corrugation limits the extent of validity of the steady Nusselt solution[9] for a liquid film flowing down an incline, because the free surface of a liquid may exhibit travelling waves. Benjamin[46] and Yih[47] found that Nusselt's solution is unstable against long–wave perturbations above a critical Reynolds number which depends on the channel inclination angle, only. Above the critical Reynolds number the flow is called convectively unstable, which means that small perturbations grow while they are travelling downstream. These findings are supplemented by the experimental work of Liu *et al.*[48] and Liu and Gollub[49]. The dynamics of single free surface waves has been studied experimentally for a free falling vertical film by Chang[12]. Vlachogiannis and Bontozoglou[50] studied the interaction of solitary waves in films flowing down flat inclines using a fluorescence imaging method. More detailed information on the instability and wave dynamics in films flowing down flat inclines has been compiled by Chang and Demekhin[51]and Craster and Matar[52].

Chapter 1. Introduction

The instability of a liquid flowing down sinusoidally undulated inclines, whose wavelength of undulation is long compared to the film thickness, has been studied analytically by Wierschem and Aksel[53] by linear stability analysis. Their finding, that the presence of a long wave undulation of the substrate tends to stabilize the flow was confirmed experimentally later by Wierschem et al.[21]. Dávalos-Orozco[54, 55] modified the Benney equation[56], which is a nonlinear evolution equation in the small wavenumber approximation, to treat flows over smoothly deformed walls and found in the particular case of a two–dimensional sinusoidal wavy wall a stabilizing influence of the walls, which gains importance with increasing steepness of the undulation. Trifonov[57, 58] applied an integral boundary layer approach to study the stability of viscous liquids flowing down corrugated surfaces over a wide range of Kapitza and Reynolds numbers and found that wall corrugations may have a stabilizing as well as a destabilizing influence on the flow. These findings have been extended and verified by Heining and Aksel[27], D'Alessio et al.[59], and Heining and Aksel[60] who studied the influence of surface tension and inertia on the stability of Newtonian and power–law liquids flowing down undulated inclines. The influence of rectangular bottom geometries on the stability of the flow has been studied experimentally by Vlachogiannis and Bontozoglou[61] and Argyriadi et al.[62] using a fluorescence imaging method. They found a remarkable stabilization of the flow at high Reynolds numbers, which proceeds through the development of a three–dimensional flow structure.

Since in most technical applications and environmental systems the flow configurations are bounded by side walls, their influence on the flow structure has to be considered. Scholle and Aksel[63] considered the effects of the liquid's capillarity at vertical channel side walls and presented an exact analytical solution of visco–capillary flow in an inclined flat channel of finite width. They find their theoretical results, which exhibit a 'velocity overshoot', already observed in early experiments by Hopf[64], to be in good agreement with their experimental data. The velocity overshoot becomes in particular pronounced in the thin film limit, which was later investigated in detail by Scholle and Aksel[65]. Furthermore, they provide a necessary condition for the flow rate, to avoid a film rupture, which is of particular interest for coating applications[6]. A detailed discussion on different competing influences of the side walls on flow rate, the velocity field and the free surface shape in the case of steady and slowly draining flow is provided by Haas et al.[66]. The role of side walls for wavefronts travelling in a channel has been studied experimentally by Leontidis et al.[67]. They find the phase velocity of the waves to be a function of the side wall distance what causes the wavefronts to exhibit parabolically curved crestlines in channel flows.

The influence of side walls on the instability of a channel flow has been studied experimentally only and the number of publications on this topic comparatively low due to the technical difficulties involved. Vlachogiannis et al.[68] studied the influence of a finite and variable channel width on the primary instability by comparing the free surface heights at two different streamwise locations by conductance probes in the small wavenumber limit. They find, that the presence of side walls has a strong stabilizing influence on the flow, when the channel is narrow and not too steep. This result has been revised by Georgantaki et al.[69] who found the Kapitza number to be a crucial parameter for the stabilizing influence of the side walls. For high Kapitza numbers they found large deviations from the stability criterion for the two–dimensional film flow. Pollak et al.[70] investigated the side wall distance and contact angle dependence of the stability experimentally. When the side wall distance is reduced to the order of the capillary length they have observed a transition of the long–wave type instability, which is typical for film flows, to a short–wave type instability, which is typical for boundary layer flows. Furthermore, the influence of capillary elevation and a velocity overshoot near the side walls on the stability of the flow is investigated.

In the present paper we discuss the impact of a two–dimensional undulation of the topography on the flow and the effects of side walls on a film flowing down a flat incline. The paper is structured as follows. In the second Chapter all the experimental systems and setups, which have been used, are presented. Chapter three deals with the flow over a two–dimensional sinusoidally undulated topography and the subtle interplay of a resonance phenomenon at the free surface and eddies appearing in the throughs of the substrae geometry. Chapter four is divided in two parts and deals with channel flow down a flat incline which is bounded by side walls. In part one the steady or basic flow is described analytically and experimentally and detailed parameter studies, like contact angle and film thickness variations are presented. The second part of Chapter four focuses on the question how the presence of side walls influences the stability of the flow. Summarizing conclusions are presented in Chapter five.

Chapter 1. Introduction

Chapter 2

Experimental systems and setups

2.1 Liquids

Three different silicone oils from *Basildon* and *Elbesil* with dynamic viscosities η ranging from approximately 10 mPas to 1000 mPas, which all showed Newtonian behavior within the considered shear rate and temperature range have been investigated. The main fluid properties at the mean measurement temperature T are summarized in Table 2.1. The density, the dynamic viscosity, the kinematic viscosity and the surface tension are denoted by ρ, η, $\nu = \eta/\rho$ and σ respectively.

Density measurements have been carried out with a Mohr Westphal balance from *Gottl. Kern & Sohn GmbH* with an absolute accuracy of $\pm 0.3 \,\text{kg/m}^3$. The temperature of the liquid in the Mohr balance was controlled by a *Lauda* thermostat type *ecoline RE204*.

Measurements of the surface tension σ were done with a *Lauda* ring–tensiometer type *TE1CA-M* whose fluid temperature was controlled by a *Lauda* thermostat type *RC 6 CP*. The resolution of the ring–tensiometer was 0.1 mN/m.

The dynamic viscosity η of the liquids has been measured with different Ubbelohde viscosimeter capillaries type *501* from *Schott*. The capillaries were plunged into a water bath whose temperature was controlled by a *Schott* thermostat within an accuracy of 0.05 °C. The precisions of the different viscosimeter capillaries were specified to be between 0.65% and 0.8%.

All fluid property measurements have been carried out in a temperature interval from 20 – 30 °C in 1 °C steps. The uncertainty of the liquid properties during an experimental run is essentially determined by the uncertainty of the liquid's temperature and was thus calculated from the temperature dependence of the liquid properties.

The temperature of the liquid flowing in the channel was measured downstream of the region of interest by *Ahlborn Mess- und Regelungstechnik GmbH* PT-100 and NTC

Manufacturer	Name	T / [°C]	ρ / [kg/m^3]	η / [mPas]	ν / [mm^2/s]	σ / [mN/m]
Basildon	BC10cs	25	924.3	10.72	11.6	18.9
Basildon	BC50	24	950.6	50.0	52.6	19.6
Elbesil	B1000	24	969.0	1,076	1,110	20.4

Table 2.1: Liquid properties of the used silicone oils.

Chapter 2. Experimental systems and setups

temperature sensors with an accuracy of 0.1 °C.

The static contact angle θ at the three–phase contact lines liquid/air/channel side wall has been measured with a contact angle goniometer from *Dataphysics* type *OCA 20*. All measured contact angles were found to be independent of the temperature within the measurement error of approximately 2° and the temperature range investigated.

2.2 Flow facilities

The experiments have been carried out in two different 170 ± 1 mm broad channels with flat bottoms made of aluminum. The side walls of channel 1 were made of Plexiglas® and channel 2 was featured with side wall clamps which allowed to mount side walls made up of different materials to vary the contact angle θ at the triple point air/liquid/side wall.

In this work we have limited our contact angle study to the two extreme cases which are technically possible. Silicone oil, which was the only fluid used throughout all experimental runs in order to keep all material parameters (see section 2.1), in particular the surface tension σ, constant, shows nearly perfect wetting characteristics on the vast majority of substrates. Because of its excellent planarity we chose Plexiglass® as a side wall material to cover this case. The static contact angle θ of silicone oil with plane Plexiglass® was measured with a sessile drop method using a static contact angle goniometer type *OCA 20* from *dataphysics* to be $8° \pm 2°$. The second set of side walls was made up of glass which has been coated with *pro.Glass® Clear 105* from *nanogate* to enlarge θ to $52° \pm 2°$ which was the largest contact angle we were able to achieve with the utilized fluid.

The overall length of channel 1 was about two meters and of channel 2 was about half a meter. The inclination angle α of both channels could have been varied continuously between 0° and 90° and has been determined by a digital protractor with an accuracy of 0.1°. The spanwise evenness of the channel has been checked by placing a water level with a display accuracy of 0.1 mm/m perpendicular to the side walls of the channel. Perpendicularity was assured with a 90°–aluminum angle placed to the side wall.

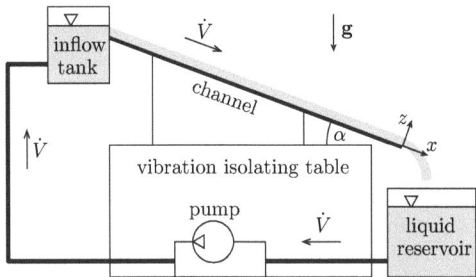

Figure 2.1: Sketch of the flow circuit including the channel which is mounted on a vibration isolating table and a pump which transports the liquid from a large liquid reservoir to a smaller inflow tank on top of the channel.

Depending on the desired flow rate $\dot{q}$ one of two different eccentric pumps from *Johstadt* provided a constant adjustable volume flux $\dot{V} = \dot{q}B$, with B being the channel width, from a large liquid reservoir into a smaller inflow tank on top of the channel. Channel 1 was equipped with a pump type *AFJ 15.1B* which provided a volume flux in the range of 1 l/min to 10 l/min. Channel 2 was equipped with a smaller pump type *AFJ 06B* which provided a volume flux in the range of 0.025 l/min to 1 l/min. Smaller volume fluxes have been realized by an adjustable bypass in the tube system between the pump and the channel. From the smaller inflow tank on top of the channel the liquid flows gravity–driven, down the channel, back into the liquid reservoir to close the flow circuit as illustrated in in Figure 2.1.

A sinusoidally undulated aluminum inlay, as illustrated in Figure 2.2, consisting of 50 periods with a wavelength $\lambda = 10\,\text{mm}$ and an amplitude $a = 1\,\text{mm}$ covering the whole width of the channel was inserted directly below the inflow of the channel 1. The gravitational acceleration is denoted by **g**, which can be written as $\mathbf{g} = (g\sin\alpha, -g\cos\alpha)$ in the (x, z)–coordinate system given in Figure 2.2.

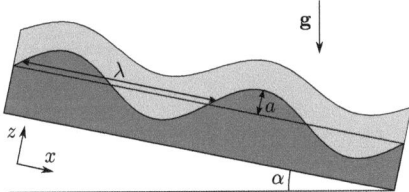

Figure 2.2: Geometry of the two–dimensional undulated inlay.

During all experimental runs the temperature of the liquid was controlled by a *TC300* thermostat from *Haake* via a heat exchanger coil sitting in the large liquid reservoir.

2.3 Tracer particles

We have used two different types of tracer particles. The mean diameter and the density of *Red Fluorescent Polymer Microspheres* from *Duke Scientifics*, which will be called fluorescent tracer particles in the following, have been specified by the manufacturer to $7\,\mu\text{m}$ and $1050\,\text{kg/m}^3$, respectively. Additionally we have determined the volume weighted particle size distribution with a *Mastersizer 2000* device from *Malvern*, which is plotted in Figure 2.3. The median particle size x_{50} and the grade of dispersity ξ_d, which is defined in[71]

$$\xi_\text{d} = \frac{x_{84} - x_{16}}{2 x_{50}}, \tag{2.1}$$

of the fluorescent tracer particles have been found to be $x_{50} = 7.122\,\mu\text{m}$ and $\xi_\text{d} = 0.3058$. The quantities x_{16}, x_{50} and x_{84} denote the particle sizes, which are greater than or equal to 16%, 50% and 84% of all particles, respectively.

The sedimentation speed u_sed of small spheres falling in a viscous liquid can be calculated to be[72]

$$u_\text{sed} = \frac{2 g r_\text{s}^2}{9 \eta}(\rho_\text{s} - \rho), \tag{2.2}$$

where r_s and ρ_s are the radius and the density of the sphere.

Chapter 2. Experimental systems and setups

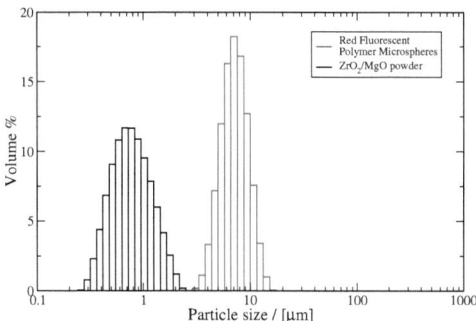

Figure 2.3: Particle size distributions.

The resulting sedimentation speeds of the fluorescent tracer particles used in the three different silicone oils BC10cs, BC50 and B1000 (see Table 2.1) were 3.2×10^{-4} mm/s, 5.5×10^{-5} mm/s and 2.1×10^{-6} mm/s, respectively. All these sedimentation velocities are at each case orders of magnitudes smaller than the typical flow velocities measured. Therefore, the sedimentation distance during one experimental run did not exceed the particles diameter.

Figure 2.4 shows the emission spectrum of the fluorescent tracer particles dissolved in silicone oil which has been measured with a *Cary Eclipse* fluorescence spectrophotometer from *Agilent Technologies* at an excitation wavelength of 532 nm.

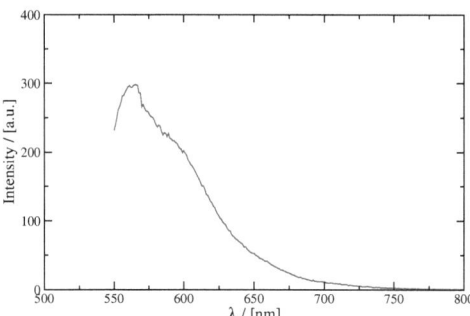

Figure 2.4: Emission spectrum of *Red Fluorescent Polymer Microspheres* from *Duke Scientifics* in silicone oil. Wavelength of the excitation light was 532 nm.

The second type of tracer particles was made up of a powder of ZrO_2/MgO from *Goodfellow*. The mean particle diameter is specified by the manufacturer to be 0.8 μm. The density is specified to be 5700 kg/m^3. The measured particle size distribution is plotted in Figure 2.3. The median particle size and the grade of dispersity have been measured to be $x_{50} = 0.76$ μm and $\xi_d = 0.4682$. These particles, which will be called scattering tracer particles in the following, were dissolved in the B1000 silicone oil from *Elbesil*, which is described in section 2.1, only. According to equation (2.2) the sedimentation speed u_{sed}

of the ZrO_2/MgO-particles in this oil was about 1.4×10^{-6} mm/s, which was orders of magnitudes smaller, than all typical velocities measured. Therefore, the sedimentation distance during one experimental run did not exceed the particles diameter.

2.4 Experimental setups

2.4.1 Flow rate

Determination of the flow rate $\dot{q}$, or the Reynolds number Re respectively, was done either by determining the film thickness d of the liquid flowing over a sufficiently long flat part in the middle of the channel or by a flow meter which measured the overall volume flux $\dot{V}$ through the channel.

Under perfectly stable flow conditions, at low Reynolds numbers, the film thickness has been measured by a micrometer screw with a needle tip. The micrometer screw was mounted to the channel in a way, that it pointed perpendicular to the free surface of the liquid or the bottom of the channel, respectively. By screwing the needle slowly towards the fluid, the position of the free surface can be detected, when the tip of the needle contacts the liquid and a capillary elevation forms instantaneously (See Figure 2.5). The position of the substrate has been determined in a similar fashion by screwing the needle further down until a small mechanical resistance was sensible. The accuracy of the film thickness determination is estimated to be better than $10 \, \mu$m.

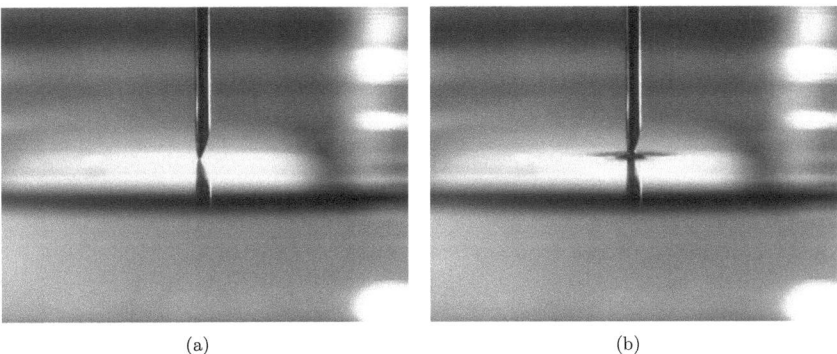

(a) (b)

Figure 2.5: Illustration of the tip of a needle which is less than 6.5 μm above the surface of a flowing liquid film (a) and just touching it (b) what causes a capillary elevation to form instantaneously. The width of the needle illustrated is 400 μm.

To determine the flow rate $\dot{q}$ at intermediate Reynolds numbers or under weakly unsteady flow conditions, the overall average volume flux $\dot{V}$ through the channel was measured by an analog flow meter which was installed between the outflow of the channel and the large liquid reservoir. For each volume flux measurement the averaging time was at least 600 s to reduce the statistical error of the volume flux measurement to less than 0.1 %.

When side wall effects are neglected and the flow is steady, the velocity profile of a

Chapter 2. Experimental systems and setups

liquid flowing down a flat incline is found to be parabolic[73]

$$u(z) = \frac{\rho g \sin \alpha}{2\eta}(2h_\mathrm{n} - z)z, \qquad (2.3)$$

where z is the cartesian coordinate perpendicular to the bottom and h_n is the film thickness of the well known Nusselt solution[9]. Integrating the velocity profile from the bottom ($z = 0$) to the free surface of the liquid ($z = h_\mathrm{n}$) yields the flow rate of the Nusselt film flow

$$\dot{q} = \int_0^{h_\mathrm{n}} u(z)\mathrm{d}z = \frac{\rho g \sin \alpha h_\mathrm{n}^3}{3\eta}. \qquad (2.4)$$

Integrating the flow rate $\dot{q}$ over the channel width B yields the volume flux $\dot{V}$.

$$\dot{V} = \dot{q}B = \frac{\rho g \sin \alpha h_\mathrm{n}^3 B}{3\eta}. \qquad (2.5)$$

When the influence of side walls on the flow is neglected, the relation between the Reynolds number Re, the film thickness of a film flowing down a flat channel and the measured volume flux is in the following given by

$$\mathrm{Re} = \frac{2u_\mathrm{s}h_\mathrm{n}}{3\nu} = \frac{\bar{u}h_\mathrm{n}}{\nu} = \frac{\dot{q}}{\nu} = \frac{\dot{V}}{\nu B}, \qquad (2.6)$$

where u_s is the free surface velocity and $\bar{u} = 2u_\mathrm{s}/3$ is the mean flow velocity.

It has to be emphasized, that equations (2.5) and (2.6) are valid, only if the velocity field u is assumed to be independent of the spanwise y-coordinate, which is only the case when the impact of the presence of side walls on the flow filed is neglected. When the influence of side walls on the flow is investigated in chapter 4, this assumption has to be dropped. A detailed discussion of the relation between the film thickness d, the measured volume flux $\dot{V}$ and the Reynolds number Re, when the influence of the presence of side walls is considered will follow in sections 4.1.2 and 4.1.3.

2.4.2 Detection of the free surface shape

In both flow facilities the shape of the free surface has been measured by illuminating fluorescent tracer particles, which are described in section 2.3, in the bulk of the liquid with a laser sheet. The fluorescent light has been detected with a charged–coupled–device (CCD) camera.

The laser sheet for channel 1 has been produced by a continuous–wave (cw) argon–Ion (Ar$^+$) Laser from *Spectra Physics* emitting at a wavelength of 514.5 nm with an approximate output power of 100 mW (See Figure 2.6). The light sheet was aligned parallel to the side walls right in the middle of the channel to study the influence of the undulated bottom on the free surface shape of the liquid flowing above it. The fluorescent light from the fluorescent tracer particles in the liquid was detected with a *JAI CV-M300* 8-bit monochrome CCD camera with a resolution of 768x494 pixels, which was inclined by about 10° with respect to the spanwise direction of the channel (See camera (2) in Figures 2.6 and 2.7). The much brighter scattered light from the underlying topography was blocked by an optical long pass filter with a 50% cut–off wavelength of 550 nm, which was mounted on a 60 mm *Nikkor micro* lens. The cameras field of view covered approximately 2.5 periods of the bottom undulation. This camera setup resulted in a spatial resolution of about 30 µm/pixel. The image was calibrated spatially with a ceramic calibration scale with a point pattern of $4\,\text{pt/mm}^2$.

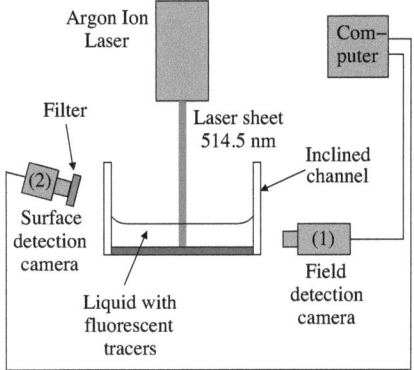

Figure 2.6: Experimental setup for the surface contour detection and the streamline detection in the troughs of the undulated inlay in channel 1. The eddy size is determined by detecting the path lines of the fluorescent tracer particles with the horizontal camera (1). The inclined camera (2) images the light sheet from the airside. The surface contour corresponds to the upper borderline of the bright sheet as seen by this camera. Reprinted with permission from [44]. ©2010, American Institute of Physics.

Figure 2.7: Photo of the experimental setup for the surface contour detection and the streamline detection in the troughs of the undulated inlay in channel 1.

The experimental setup for the free surface detection in channel 2 is sketched in Figure 2.8. The laser sheet has been aligned perpendicular to the side walls of the channel and was expanded to illuminate the liquid in a region from the side wall to approximately 50 mm apart from it, to detect the shape of the capillary elevation of the liquid in the proximity of the side walls. The sheet was produced by a frequency doubled, pulsed

Chapter 2. Experimental systems and setups

neodymium-doped yttrium aluminum garnet (Nd:Yag) laser from *New wave research* type *Solo II 15Hz* emitting at a wavelength of 532 nm. The pulse length and energy is specified to be 6 ns and 100 mJ. An optical device from *Cosmicar/Pentax* was attached directly to the laser head to create the light sheet which had a width of approximately 100 μm. The scattered primary light from the bottom was blocked in front of the camera's lens using the same long pass filter as described above. The fluorescent light from the fluorescent tracer particles was detected by a monochrome CCD camera *HiSense* from *Dantec* with a resolution of 1280x1024 and a capturing rate of 8 Hz. The camera was inclined by about 15° with respect to the channel inclination as illustrated in Figure 2.8. The spatial resolution of this camera setup was about 8 μm/pixel. Calibration of the image has been carried out *a priori* by placing a millimeter scale at the laser sheet position. Camera and laser have been synchronized by a triggering unit from *Dantec*.

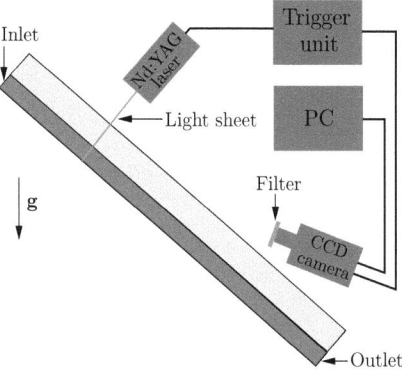

Figure 2.8: Measurement setup for the free surface shape detection. Reprinted with permission from [66]. ©2011, American Institute of Physics.

Capturing images from the liquid as described above resulted in grainy single images, because the fluorescent tracer particles sit at discrete points when one image is taken. Therefore, it was necessary to average or superimpose several images to get a uniformly bright sheet in the image whose upper border corresponds to the free surface contour of the liquid at the position of the laser sheet. Depending on the case the post–processing workflow differs slightly. Therefore, a more detailed description about the method of how the averaging or superimposing of the images was performed will be given where the corresponding results are presented.

2.4.3 Streamline detection

A detection of the streamline pattern of the liquid in the throughs of the undulated inlay in channel 1 has been done in a similar way as described by Wierschem *et al.*[33, 35, 44]. The fluorescent tracer particles and light sheet are identical with the ones described in subsection 2.3 and 2.4.2. The scattered light from the illuminated particles was detected with a frame rate of up to 500 Hz depending on the mean flow velocity with a monochrome high–speed camera *CamRecord 600* from *Optronics*. The camera was aligned perpendicular to the channel side walls as illustrated in Figures 2.6 and 2.7 (camera (1)). The streamline patterns have been reconstructed by superposing 2048 single images taken in one run with the camera's full resolution of 1280x1024 pixel by taking at each pixel position the brightest pixel of all images (see Figure 2.9(b)). The contour and position of the underlying topography could have been reconstructed in a similar fashion, but by taking not the brightest but the darkest pixel of all pictures at each pixel position (see Figure 2.9(a)). The bright lines below contour line of the underlying topography in Figure 2.9(b) came from reflections at the aluminum substrate.

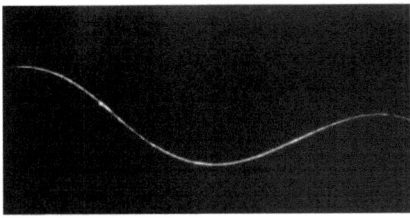 

(a) Contour of the underlying topography reconstructed by taking the darkest pixel of a 2048 series images at each pixel position.

(b) Streamline pattern reconstructed by taking the brightest pixel of a 2048 images series at each pixel position.

Figure 2.9: Illustration of the evaluation method for the reconstruction of the shape of the underlying topography and the streamline pattern from experimental single image data.

The resulting spatial resolution was about $12\,\mu$m/pixel. Spatial calibration of the images has been carried out with help of the ceramic calibration scale as described in subsection 2.4.2.

2.4.4 Velocity field measurements

Velocity measurements have been done with a *Laser-Doppler-Velocimeter* (LDV) from *Dantec/Invent*. A detailed assessment of the general accuracy of the LDV-measurement technique when it is applied on film flows is provided by Aksel and Schmidtchen[74]. As tracer particles a powder of ZrO_2/MgO with a mean particle diameter of $0.8\,\mu$m, which is described more detailed in section 2.3, was used.

The light source of the LDV–system was a *Spectra Physics* Argon–Ion (Ar^+) Laser emitting light at three main wavelengths of 476.5 nm, 488 nm and 514.5 nm. A *Dantec FiberFlow* beam splitter divided the laser beam into two equally intense beams and coupled the three colors into 6 glass fibre optics. Additionally, a Bragg cell shifted one of the two laser beams by 40 MHz to higher frequencies before the beams are splitted into their different wavelengths for two reasons. One, to generate heterodyne detection signals with a sufficiently high frequency from slow scattering tracer particles and two, to obtain information about the direction of the flow. Because the LDV system has been used in the one–dimensional (1D) mode only, all wavelengths except for the most intense (514.5 nm) were blocked by mechanical shutters.

An optical device (LDV-head) from *Dantec* with an extra focussing unit crossed the two remaining working frequency laser beams in an elliptical measurement volume which was specified to be $25\,\mu$m $\times$ $24\,\mu$m $\times$ $126\,\mu$m in size.

Chapter 2. Experimental systems and setups

The LDV-head was mounted on a X-Y-Z-traverse from *Isel* to position the LDV–measurement volume in the liquid flowing down the inclined channel. The plane in which the two laser beams crossed was always aligned parallel to the inclination plane of the channel, therefore the measured velocity was identical with the streamwise velocity component u. The step width of the traverse is specified to be 12.5 μm in X- and Y-direction and 6.25 μm in Z-direction.

2.4.5 Stability measurements

The measurements on the free surface stability of a liquid flowing down a flat channel of finite width have been carried out in the flow facility of channel 2 which is described in section 2.2. The inclination angle has been kept constant throughout all stability measurement runs at $40.8 \pm 0.05°$ to provide good comparability to the work of Haas et al.[66]. As liquid silicone oil *Basildon BC50*, which is characterized in section 2.1, has been used. To study the influence of the contact angle θ between the liquid and the side walls of the channel two different sets of side walls have been used which are described in section 2.2.

A film flow is called linear convectively unstable at a certain perturbation frequency, when an infinitesimal small free surface wave of the corresponding wavelength is growing in size while it is travelling downstream, otherwise, when a wave is damped while it is travelling downstream the flow is called convectively stable.[75, 76]

Figures 2.10 and 2.11 show the experimental setup, which is similar to the ones described by Liu et al.[48], Wierschem et al.[21] and Pollak et al.[70], to study the stability of the flow configuration described above. An eccentric pump from *Johstadt* provided an adjustable constant volume flux $\dot{V}$ which was perturbed sinusoidally in time by a 169 mm broad paddle which oscillated, driven by a stepping motor, with an amplitude of about 100 μm within the liquid in the small inflow tank on top of the channel. The excitation frequency f_e could have been chosen continuously in the range of 0.8 Hz to more than 10 Hz.

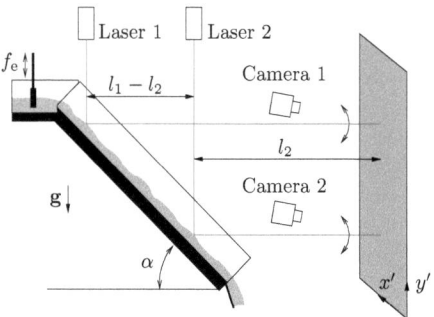

Figure 2.10: Sketch of the experimental setup for the free surface stability measurements. Reprinted with permission from [70]. ©2011, American Institute of Physics.

To detect whether the generated waves grow or decay in size while travelling downstream two laser beams generated by two identical cw Helium-Neon (HeNe) lasers from

2.4. Experimental setups

Figure 2.11: Photo of the experimental setup for the measurements of the free surface stability.

JDS Uniphase with an output power of 1 mW were alinged parallel and pointed at the free surface of the liquid. The reflection point of the upper laser (Laser 1 in Figure 2.10) hit the liquid's free surface approximately 15 cm below the inflow of the channel, the lower laser (Laser 2) hit the free surface approximately 10 cm above the outflow of the channel, to avoid inflow and outflow disturbances. The travel distances l_1 and l_2 of the laser beams from their reflection point to the screen were $l_1 = 167$ cm and $l_2 = 149.5$ cm. The distance d_s to the side walls (see Figure 2.12) was variable and could have been adjusted with an accuracy of about 0.25 mm, which is approximately half of the beam diameter. The inclination of the laser beams was 8.4° with respect to the direction of gravity, chosen in a way that the undisturbed reflected laser beams traveled horizontally to the screen. According to the additional slope of the liquid's free surface which is generated by a wave passing the reflection point of the laser, the laser spot on the screen gets deflected by an amplitude which is proportional to the wave height.

During each experimental run each of two identical *JAI CV-M10BX* 8-bit monochrome CCD cameras with a resolution of 782x582 pixels captured 512 images of one laser spot on the screen with a rate of 25 frames per second (fps). After applying Gaussian filters to the images to reduce grain the laser spot position **p** was determined in each single image

Chapter 2. Experimental systems and setups

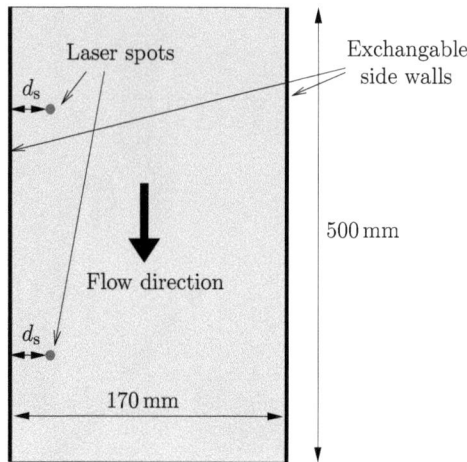

Figure 2.12: Channel top view illustrating the measurement positions. Reprinted with permission from [70]. ©2011, American Institute of Physics.

by calculating the center of area of a minimum threshold grey–value area. The threshold usually was set to 80% of the brightest pixel in each image after the Gaussian filters were applied.

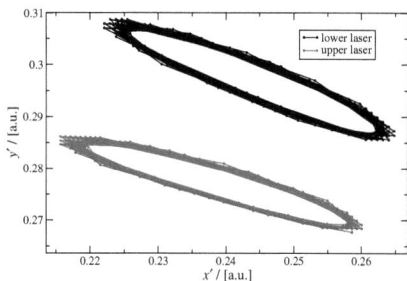

Figure 2.13: Recorded positions of both laser spots during an experimental run. $d_s = 20$ mm, $f_e = 3.2$ Hz, Re $= 2.301$, and $\theta = 8°$. Reprinted with permission from [70]. ©2011, American Institute of Physics.

Typical path lines of the resulting laser spot positions $\mathbf{p}_i(t) = (p_{x',i}(t), p_{y',i}(t))$, which have been rescaled with respect to the geometry of the setup, are plotted in Figure 2.13. The variables x' and y' denote the cartesian coordinates of a coordinate system located on the screen as indicated in Figure 2.10. The elliptical movement of the two laser spots has always been observed when off–center measurements were made, because the wave fronts of travelling free surface waves in a channel of finite width are not straight[49, 67].

Figure 2.14 shows a small section of the time dependence of the x'- and y'-components of the laser spot path lines illustrated in Figure 2.13. The shape of these curves already

2.4. Experimental setups

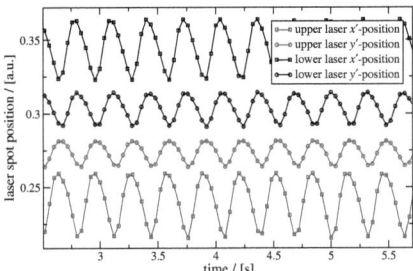

Figure 2.14: Section of the time dependence of the x'- and y'-components of both laser spots during an experimental run. Signals have been shifted vertically to avoid overlapping. $d_s = 20\,\text{mm}$, $f_e = 3.2\,\text{Hz}$, $\text{Re} = 2.301$, and $\theta = 8°$. Reprinted with permission from [70]. ©2011, American Institute of Physics.

indicates, that the signal mainly consists of a sinusoidal function with a single frequency which matches the excitation frequency f_e.

To quantify the amplitudes of the laser spot oscillations, fast Fourier transformations (FFT) have been applied to the signals $\mathbf{p}_i(t)$. Figures 2.15(a)-2.15(c) show typical absolute values

$$\hat{p}_i(f) = \sqrt{\hat{p}_{x',i}^2(f) + \hat{p}_{y',i}^2(f)}$$

of the Fourier transformed signals

$$\hat{\mathbf{p}}_i(f) = \left(\hat{p}_{x',i}(f), \hat{p}_{y',i}(f)\right) = \left(\mathcal{F}(p_{x',i}(t)), \mathcal{F}(p_{y',i}(t))\right).$$

One can clearly identify a dominant peak at the fundamental excitation frequency and one[1] much smaller peak at higher order, showing that the signals $\mathbf{p}_i(t)$ are mainly sinusoidal with a fundamental frequency which corresponds to the frequency excited by the paddle f_e as already suspected from Figure 2.14.

To compare the amplitudes of the two laser spots on the screen, and therefore the amplitudes of the travelling free surface waves at the up- and the downstream point, the fundamental peak of the Fourier transformed signal $\hat{p}(f)$ is fitted with a Gaussian function. The height difference of the Gaussian fit functions of the peak from the Fourier transformed signals from the upper and the lower laser $\Delta\hat{p}_{12}$ is plotted versus the Reynolds number Re in Figure 2.15(d).

The neutral point at which the free surface waves are neither damped, nor amplified has been determined by a linear fit of the of the data very close to the neutral point only, as illustrated in Figure 2.15(d).

To provide comparability of our experimental results to the theory of small perturbations, we estimate the maximal amplitude of the generated free surface waves in the following. Therefor, we assume that the shape of a travelling free surface wave can be described by a single sinusoidal function, as proven above. The maximal slope $m_{\max}$ of such a wave is given by

$$m_{\max} = \frac{2\pi A}{\lambda_{\text{w}}}, \tag{2.7}$$

---
[1] For smaller excitation frequencies also more.

Chapter 2. Experimental systems and setups

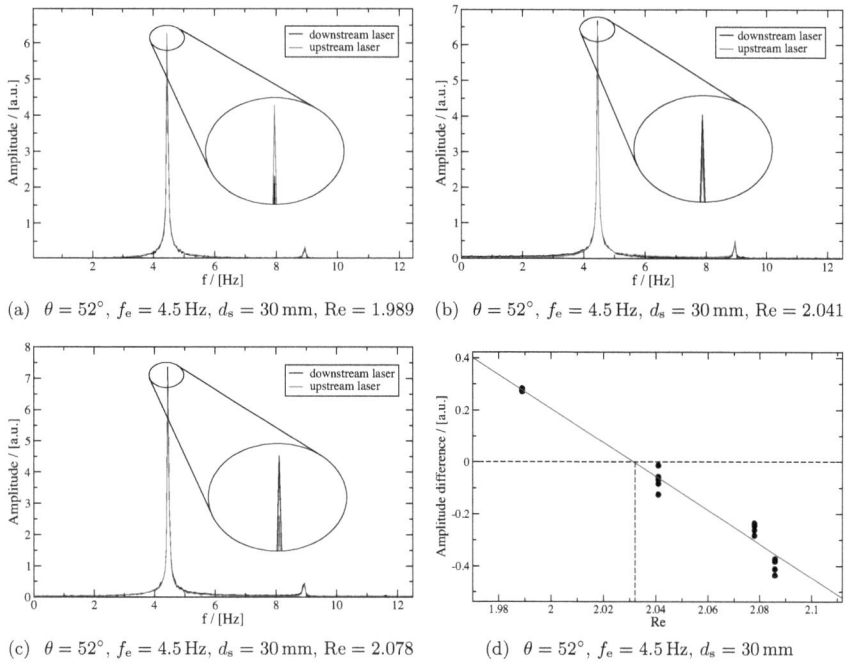

(a) $\theta = 52°$, $f_e = 4.5\,\text{Hz}$, $d_s = 30\,\text{mm}$, Re = 1.989

(b) $\theta = 52°$, $f_e = 4.5\,\text{Hz}$, $d_s = 30\,\text{mm}$, Re = 2.041

(c) $\theta = 52°$, $f_e = 4.5\,\text{Hz}$, $d_s = 30\,\text{mm}$, Re = 2.078

(d) $\theta = 52°$, $f_e = 4.5\,\text{Hz}$, $d_s = 30\,\text{mm}$

Figure 2.15: Absolute values of the Fourier transformed signals of both lasers at different Reynolds numbers and their amplitude differences. Each diagram (a-c) shows the spectrum of a single experimental run. Diagram (d) shows the dependence of $\Delta\hat{p}_{12}$ on the Reynolds number. Reprinted with permission from [70]. ©2011, American Institute of Physics.

where A and λ_w are its amplitude and wavelength, respectively. The largest amplitudes A result from large laser beam deflections which are generated by long waves (see equation (2.7)). Thus, we approximate the maximal wavelength throughout all our stability measurement runs by

$$\lambda_\text{max} = \frac{u_\text{s,max}(f_\text{min})}{f_\text{min}} = \frac{3\dot{V}_\text{max}(f_\text{min})}{2Bh_\text{max}(f_\text{min})f_\text{min}} = \frac{1}{\sqrt{2}f_\text{min}}\sqrt[3]{\text{Re}_\text{max}(f_\text{min})^2 \nu \sin\alpha g}, \quad (2.8)$$

with $u_\text{s,max}$, h_max, f_min, $\dot{V}_\text{max}$ and Re_max beeing the maximal free surface velocity, the maximal film height, the minimal excitation frequency, the maximal volume flux and the maximal Reynolds number throughout all experimental stability measurement runs, respectively. The maximal slope of a travelling wave can also be estimated geometrically from the amplitude of the deflection of a laser beam δp_i on the screen by

$$m_\text{max} \approx \delta p_1/l_1 = \delta p_2/l_2. \quad (2.9)$$

This approximation is reasonable when the impact of the shifting of the reflection point at the free surface is negligible compared to the impact of the additional slope of the free surface on the deflection of the laser beam, which is certainly the case here because of the long travel distances l_1 and l_2 of the laser beams. Since the amplitudes δp_i on the screen did never exceed 1 cm at any excitation frequency, we conclude, that the amplitude A of the generated waves at the free surface did never exceed a value of $52\,\mu$m.

Chapter 2. Experimental systems and setups

Chapter 3

Two–dimensional film flow

3.1 Suppression of eddies

3.1.1 Problem formulation

We consider a steady two–dimensional gravity–driven flow of a Newtonian liquid down an inclined topography which is sinusoidally undulated in the main flow direction. The profile of the substrate undulation which is described by

$$b(x) = a\cos(2\pi x/\lambda), \tag{3.1}$$

is illustrated in Figure 3.1, where a is the amplitude and λ the wavelength of the periodic undulation. The gravitational acceleration $\mathbf{g}$ is given by $\mathbf{g} = g(\sin\alpha, -\cos\alpha)$ in the given

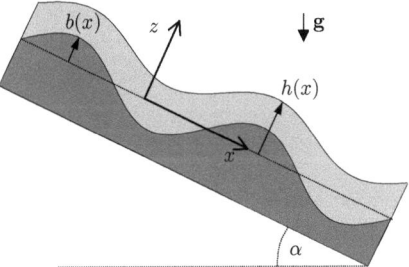

Figure 3.1: Viscous film flow down a wavy incline. Reprinted with permission from [44]. ©2010, American Institute of Physics.

(x, z)-coordinate system, with α beeing the mean inclination angle of the channel. The position of the liquid's free surface is denoted by $h(x)$. The film thickness d can easily be calculated by substracting the bottom contour $b(x)$ from the free surface shape $h(x)$:

$$d(x) = h(x) - b(x). \tag{3.2}$$

The Navier–Stokes equations and the continuity equation for incompressible liquids

$$\rho(\mathbf{u}\cdot\nabla)\mathbf{u} = -\nabla p + \eta\Delta\mathbf{u} + \rho\mathbf{g}, \quad \nabla\cdot\mathbf{u} = 0 \tag{3.3}$$

are rewritten in a dimensionless form by introducing reference quantities. As a characteristic length we use the film thickness d_n of the corresponding flow over a flat incline with the same flow rate $\dot{q}$. From the well known Nusselt solution[73] it can be calculated to

$$d_\mathrm{n} = \sqrt[3]{3\nu\dot{q}/(g\sin\alpha)}. \tag{3.4}$$

Consequently, velocities are rescaled with the mean velocity of the corresponding Nusselt film flow which reads

$$\bar{u}_\mathrm{n} = \dot{q}/d_\mathrm{n} = (gd_\mathrm{n}^2 \sin\alpha)/(3\nu). \tag{3.5}$$

The pressure is rescaled with the dynamic pressure $\rho\bar{u}_\mathrm{n}^2$. Inserting these scalings into (3.3) yields a dimensionless formulation of the Navier–Stokes and the continuity equations

$$\mathrm{Re}\,(\tilde{\mathbf{u}}\cdot\nabla)\tilde{\mathbf{u}} = -\mathrm{Re}\nabla\tilde{p} + \Delta\tilde{\mathbf{u}} + \tilde{\mathbf{g}}, \quad \nabla\cdot\tilde{\mathbf{u}} = 0, \tag{3.6}$$

with the Reynolds number

$$\mathrm{Re} = \bar{u}_\mathrm{n} d_\mathrm{n}/\nu = \dot{q}/\nu \tag{3.7}$$

and the dimensionless gravity vector $\tilde{\mathbf{g}} = (3, -3\cot\alpha)$, where a tilde $\tilde{}$ denotes a dimensionless quantity. The shape of the underlying topography is described by the dimensionless steepness parameter $\xi = a/d$ and the dimensionless wave number $k = 2\pi d/\lambda$.

At the bottom of the topography $\tilde{y} = \tilde{b}(\tilde{x}) = \xi\cos(k\tilde{x})$ the no–slip condition $\tilde{\mathbf{u}} = \mathbf{0}$ holds. Because the liquids free surface contour is a streamline the kinematic boundary condition

$$\frac{\mathrm{d}\tilde{h}}{\mathrm{d}\tilde{x}} = \frac{\tilde{v}}{\tilde{u}} \tag{3.8}$$

has to be fulfilled at $\tilde{y} = \tilde{h}$, where $\tilde{u}$ and $\tilde{v}$ are the velocity components in $\tilde{x}$ and $\tilde{y}$ direction, respectively. The dynamic boundary condition, which takes care of the balance of stresses at the free surface, has to be fulfilled and reads

$$\mathbf{n}\cdot\underline{\mathbf{T}} = \left(\frac{3\mathrm{Bo}^{-1}}{k^2\mathrm{Re}}\right)\kappa\mathbf{n} \tag{3.9}$$

when the viscosity of air is neglected. The outer normal unit vector of the free surface is denoted by $\mathbf{n}$, $\underline{\mathbf{T}}$ is the stress tensor $\underline{\mathbf{T}} = -(\tilde{p} - \tilde{p}_0)\underline{\mathbf{I}} + (1/\mathrm{Re})[\nabla\tilde{\mathbf{u}} + (\nabla\tilde{\mathbf{u}})^T]$, $\underline{\mathbf{I}}$ is the identity matrix, p_0 is the ambient pressure and κ is the curvature of the free surface which is given by

$$\kappa = \frac{1}{R} = \frac{\frac{\mathrm{d}^2 h}{\mathrm{d}x^2}}{\left[1 + \left(\frac{\mathrm{d}h}{\mathrm{d}x}\right)^2\right]^{3/2}}, \tag{3.10}$$

with R being the radius of curvature of the free surface shape. The inverse Bond number $\mathrm{Bo}^{-1} = 4\pi^2\sigma/(\rho g\lambda^2 \sin\alpha)$ is a measure for the ratio between surface tension stresses and gravitational stresses. We have now formulated the problem (3.3) with the given boundary conditions in a dimensionless form in a way that it is governed by five independent dimensionless parameters namely ξ, k, Bo^{-1}, Re and $\cot\alpha$, only.

Additionally, we claim the flow to be periodic as the underlying topography is. Therefore, we assume the free surface shape, the pressure and the velocity field to be periodic in downstream- ($\tilde{x}$-) direction:

$$\tilde{h}(\tilde{x}) = \tilde{h}(\tilde{x} + 2\pi/k) \qquad (3.11)$$
$$\tilde{p}(\tilde{x}, \tilde{y}) = \tilde{p}(\tilde{x} + 2\pi/k, \tilde{y}) \qquad (3.12)$$
$$\tilde{\mathbf{u}}(\tilde{x}, \tilde{y}) = \tilde{\mathbf{u}}(\tilde{x} + 2\pi/k, \tilde{y}). \qquad (3.13)$$

With the kinematic (3.8) and the dynamic (3.9) boundary conditions at the free surface and the no–slip boundary condition at the bottom, the periodic boundary conditions (3.11)-(3.13) we complete the set of dimensionless field equations (3.6).

The fact that the position of the free surface, and therefore also the domain of solution, is not known *a priori* introduces a further degree of freedom, which has to be captured by a numerical procedure. This has been achieved by implementing an iterative procedure starting with an initial guess for the free surface shape. In each solution step the Navier–Stokes equations and the continuity equation (3.6) have been solved together with the no–slip condition at the substrate $\mathbf{u} = \mathbf{0}$ and the dynamic boundary condition at the free surface (3.9). The kinematic boundary condition (3.8) cannot be fulfilled yet and is formally interpreted as a first order differential equation for the unknown free surface position $\tilde{h}(\tilde{x})$. With the new free surface position the iterative procedure is repeated until the difference between the solution of the previous and the current step is below a threshold value. Since the flow rate is still arbitrary we claimed a certain flow rate and therefore a certain Reynolds number to obtain a unique solution.

In each iteration step, the velocities and the pressure are approximated using the Taylor–Hood element pair with piecewise quadratic velocity approximation and piecewise linear pressure. The resulting nonlinear equation for the nodal velocity and pressure is solved with Newton method, which typically converges in four to six iterations.

The numerical procedure described above has been implemented and all numerical calculations presented in this chapter have been carried out by Christian Heining[44].

3.1.2 Experimental and numerical findings

The experiments have been carried out with *Basildon* silicone oil BC10cs, which is described in section 2.1, flowing over a sinusoidally undulated inlay with an amplitude $a = 1\,\text{mm}$ and a wavelength $\lambda = 10\,\text{mm}$ placed in channel 1 near its inflow as described in section 2.2. Measurements have been done at four different inclination angles ranging from $5°$ to $14°$. The Reynolds number has been varied between 3 and 63.

Streamline patterns have been recorded experimentally as described in section 2.4.3. Figure 3.2 shows a comparison of experimentally and numerically observed streamlines at different flow rates or Reynolds numbers, respectively. The bright sinusoidal line which is overlaid by a red one corresponds to the substrate geometry. The lines above are the numerically (green) and experimentally (black & white) determined streamlines. The uppermost red line corresponds to the numerically determined free surface contour. The bright lines below the bottom and the inversely bent lines above the free surface are reflections coming from the substrate or the free surface of the liquid.

We find that the flow shows qualitatively different behavior depending on the flow rate or the Reynolds number, respectively. When the mean film thickness d is small compared to the wavelength λ and the amplitude a the flow can locally be well described by the Nusselt solution with the local inclination angle[21]. Therefore, the liquid flows

Chapter 3. Two–dimensional film flow

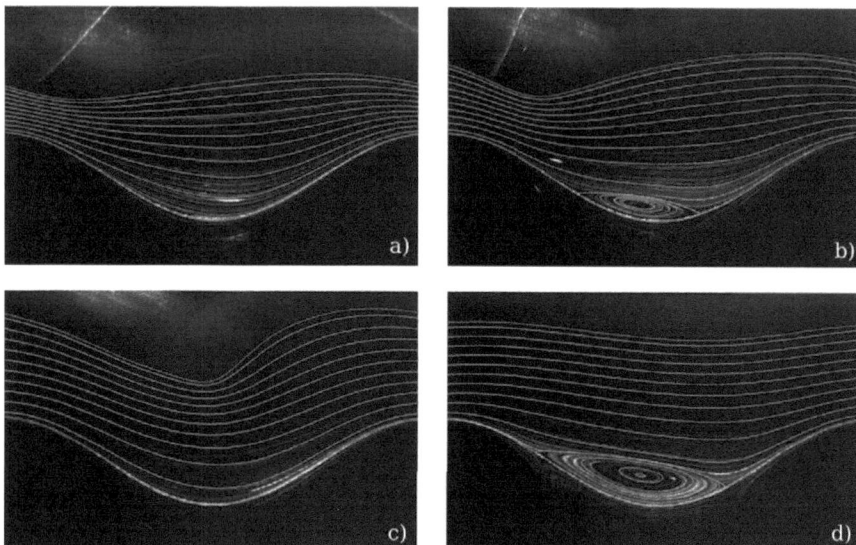

Figure 3.2: Comparison of experimental path lines to numerical streamlines. The images are rotated by the mean inclination angle of the channel. The volume flux is continuously increased form a) to d). a) Re = 9: no eddy at low Reynolds numbers. b) Re = 16: increasing inertia results in the generation of an eddy in the trough of the undulation. c) Re = 31: increasing inertia further, the eddy vanishes. d) Re = 48: flow separation reappears at even higher Reynolds numbers. Bottom contour: lower bright sinusoidal line; lines below and inversely bent lines in the upper part of the images are reflections of the path lines at the bottom and at the free surface. Channel inclination angle $\alpha = 8°$. Reprinted with permission from [44]. ©2010, American Institute of Physics.

smoothly along the substrate contour when the Reynolds numbers are rather small (See Figure 3.2a)). When the flow rate is increased the flow begins to separate into a region where the flow recirculates within the trough of the undulation and into a main flow above (See Figure 3.2b)). In contrast to the vast majority of systems where an increase of the influence of inertia leads to a growth of recirculation areas[77], increasing the Reynolds number in this system leads to diminution of the eddies until they vanish completely as shown in Figure 3.2c). However, increasing the Reynolds number further, we find a critical Reynolds number at which the eddies reappear as depicted in Figure 3.2d). Above this critical Reynolds number the eddies grow monotonously in size with increasing Reynolds number. All streamline patterns shown in Figure 3.2 correspond to a mean channel inclination angle α of $8°$. Qualitatively similar results have been obtained for other inclination angles. We remark that closed path lines, the lack of jitter and the excellent agreement between experiment and numerics indicate that unsteady motion is negligible and that the periodicity (3.11)-(3.13) and two–dimensionality assumptions made in section 3.1.1 hold.

Detection of the free surface shape of the liquid has been done as described in section 2.4.2. Because the fluorescent tracer particles described in section 2.3 sit at discrete

3.1. Suppression of eddies

points when an image is taken, a single recording resulted in a grainy image of the liquid (see Figure 3.3(a)). Thus, the evaluation of the free surface shape has been done with an image, which has been averaged over 50 single images (See Figure 3.3(b)). We note, that the average film thickness in Figure 3.3 appears much thinner as it is because the image of the liquid below the free surface (and also of the underlying topography) is strongly distorted by the curved surface of the liquid.

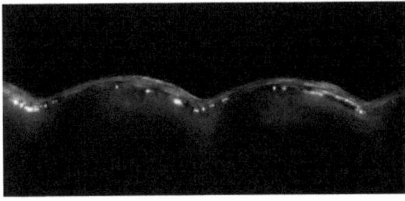 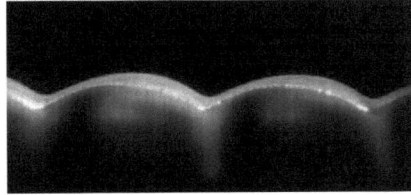

(a) Single image from surface detection camera (2).   (b) Image averaged over 50 single recordings.

Figure 3.3: Illustration of the averaging process for the detection of the free surface shape. The upper border of the bright sheet in each image corresponds to the contour of the free surface.

As Figure 3.2 indicates not only the flow in the troughs of the undulation shows a strong Reynolds number dependence but also the free surface shape of the liquid changes considerably with the Reynolds number. Figure 3.4 provides a comparison of measured (symbols) and calculated (lines) free surface shapes at different Reynolds numbers for a mean channel inclination angle α of $8°$. The curves are staggered in z-direction to avoid overlapping. The Reynolds number ranges from 6.6 (lowermost curve) to 56.2 (uppermost curve). At low Reynolds numbers we find the free surface shape to be rather harmonic and of small amplitude. When Re is increased the free surface rapidly gains in amplitude, is shifted downstream and exhibits a sharp nonlinear indentation in the trough which becomes maximal at Re = 31.5 (See red 'x'-symbols in Figure 3.2). Increasing the Reynolds number just a little further, from Re = 31.5 to Re = 32.1 in Figure 3.2, causes this indentation to vanish very brusquely resulting in a smooth sinusoidal shape again (See green '+'-symbols in Figure 3.2). Further increase of Re shifts the free surface contour further downstream while its amplitude decreases continuously. Qualitatively similar results have been obtained for other inclination angles.

Recirculation areas

Based on the experimentally and numerically obtained streamline patterns we have evaluated the size of the recirculation areas in the troughs of the undulation. Figure 3.5 shows the eddy area as a function of the Reynolds number at four different mean inclination angles α. Except for the steepest inclination angle of $14°$ eddies appear with increasing Reynolds number at a first critical Reynolds number $Re_1 \approx 11$ which seems to be independent of the channel inclination angle α. Then the eddy size increases until it reaches a local maximum and shrinks again until it vanishes completely at a second critical Reynolds number Re_2. Only beyond a third critical Reynolds number Re_3 the eddies reappear and grow monotonously in size with increasing Reynolds number in the investigated range. While the second critical Reynolds number for the disappearance of

Chapter 3. Two–dimensional film flow

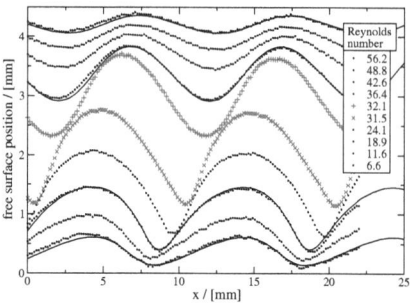

Figure 3.4: Comparison of experimental and numerical free surface shapes at different Reynolds numbers. Experimental data are represented by symbols; numerical data are represented by lines. The free surface positions are shifted perpendicular to the mean flow direction to avoid overlapping. The vertical position augments with Reynolds number. Channel inclination angle $\alpha = 8\,^\circ$. Reprinted with permission from [44]. ©2010, American Institute of Physics.

the recirculation areas showed a strong channel inclination angle dependence, the critical Reynolds numbers Re_1 and Re_3 for the emerging of eddies seemed to be rather independent of α in the investigated inclination angle range.

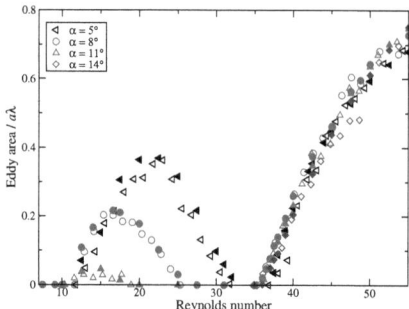

Figure 3.5: Cross–sectional area of the eddy as a function of the Reynolds number at different inclination angles. Experimental and numerical data are represented by open and solid symbols, respectively. At least 40 measurements per inclination angle have been carried out from $Re \approx 6$ to $Re \approx 62$ in equidistant steps. Where no eddy was observed, most data points have been blanked out for clarity. Reprinted with permission from [44]. ©2010, American Institute of Physics.

Thus, we find that eddies which appear at not too steep inclination angles at a critical Reynolds number Re_1 disappear again in an eddy–free window whose extent grows with increasing α. The window where eddies can be observed between Re_1 and Re_2 shrinks accordingly at the expense of the eddy–free window until it vanishes completely for an inclination angle $\alpha = 14\,^\circ$.

3.1. Suppression of eddies

Free surface shape

Quantitative analysis of the free surface shape data shown in Figure 3.4 has been done by decomposing them into Fourier series by discrete Fourier transformation (DFT). Figures 3.6-3.8 show the Reynolds number dependence of the amplitudes of the zeroth, the first and the second Fourier modes for all inclination angles studied.

The zeroth Fourier mode is illustrated in Figure 3.6 and corresponds to the film height[1] $h(x)$ averaged over one period of the bottom contour. At Reynolds numbers below ≈ 25 and above ≈ 37 we find a monotonous increase of the average film thickness with increasing volume flux or Reynolds number, respectively for all investigated channel inclination angles, as it is common for gravity–driven film flows[33, 34, 35]. In the region at intermediate Reynolds numbers all data sets reveal a spontaneous drop in the average film thickness with increasing Reynolds number. The position where this drop takes place shifts with increasing channel inclination to smaller Reynolds numbers.

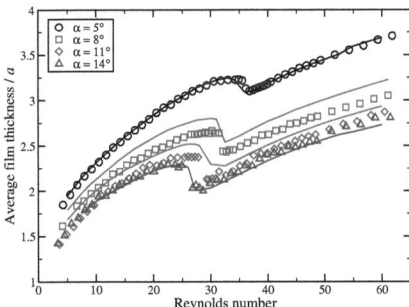

Figure 3.6: Mean film thickness averaged over one bottom period as a function of the Reynolds number at different inclination angles. Experimental and numerical data are represented by symbols and lines, respectively. Reprinted with permission from [44]. ©2010, American Institute of Physics.

The first harmonic of the Fourier transformed free surface shape corresponds to the wavelength of the bottom contour. Its amplitude shows a peak which grows and shifts its position from Re ≈ 35 to Re ≈ 27 to smaller Reynolds numbers when the channel inclination angle α becomes steeper (See Figure 3.7). The presence of this peak reflects the fact that the free surface is strong undulated where the first harmonic peaks, but rather flat for low and for high Reynolds numbers as already illustrated in Figure 3.4.

The amplitude of the second harmonic of the free surface shape characterizes its nonlinearity. We find that it grows with increasing Reynolds number and reaches a plateau before it drops discontinuously to a much smaller value and tends rapidly against zero for large Reynolds numbers, as visible from Figure 3.8. The growth and rapid drop of the second Fourier mode corresponds to the emerging of the sharp indentation in the troughs of the free surface shape, as can be seen most clearly from the red line in Figure 3.4, and the abrupt shape transition to a smooth sinusoidal one as represented exemplarily by the green line in Figure 3.4. The height of the plateau grows with steeper channel inclinations.

---
[1] or film thickness $d(x) = h(x) - b(x)$.

Chapter 3. Two–dimensional film flow

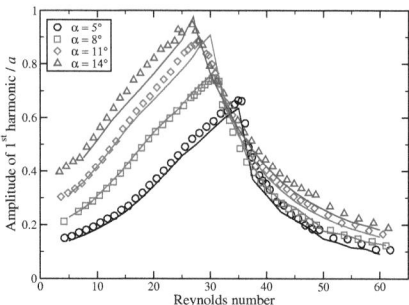

Figure 3.7: Amplitude of the first harmonic of the free surface shape as a function of the Reynolds number at different inclination angles. Experimental and numerical data are represented by symbols and lines, respectively. Reprinted with permission from [44]. ©2010, American Institute of Physics.

In conjunction to the growth, the position of the sharp drop is shifted to smaller Reynolds numbers with steepening the mean channel inclination. All higher harmonics, which are not shown here, showed qualitatively the same behavior as the second one but with much smaller amplitudes.

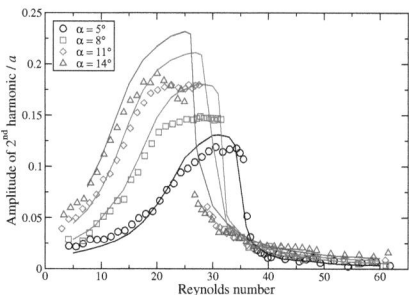

Figure 3.8: Amplitude of the second harmonic of the free surface shape as a function of the Reynolds number at different inclination angles. Experimental and numerical data are represented by symbols and lines, respectively. Reprinted with permission from [44]. ©2010, American Institute of Physics.

A comparison of Figures 3.6, 3.7 and 3.8 reveals that the drop of the average film thickness, the peak position of the first Fourier mode and the drop of the amplitude of the second Fourier mode seem to coincide at the same Reynolds number for each channel inclination angle. Figure 3.9 complies the first harmonics for the four investigated inclination angles. We find that for all investigated channel inclinations the first harmonic peaks, where the average film thickness drops and the shape of the free surface undergoes a sharp transition from an anharmonic shape to a smooth sinusoidal one. This position where a surface shape transition occurs is indicated by the dashed line in each diagram. The Reynolds number where this transition takes place shifts with steeper channel inclinations to smaller Reynolds numbers.

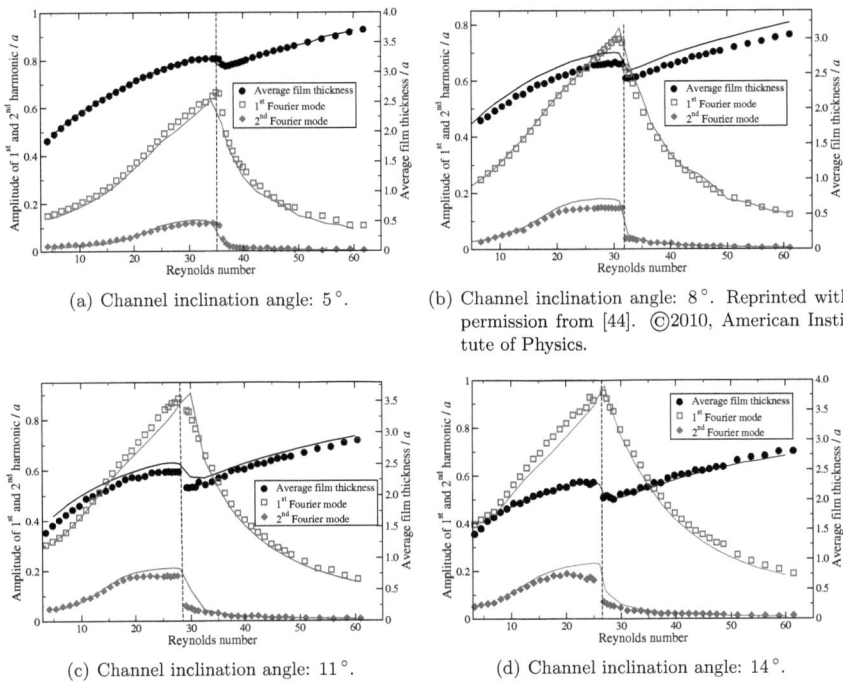

Figure 3.9: Average film thickness and amplitude of the first two Fourier components. The transition Reynolds number is indicated by the dashed line. The experimental and numerical data are represented by symbols and lines, respectively.

Figure 3.10 shows a snapshot of a video[2], which illustrates the transition of the liquid's free surface shape while the Reynolds number has been increased continuously over the transition Reynolds number from Re ≈ 26 to Re ≈ 34 from two different perspectives at a fixed channel inclination angle of $\alpha = 8\,°$. The main frame depicts a slightly deformed picture of two and a half periods of the free surface shape. The main flow direction in the main frame is from right to left. Starting from low Reynolds numbers (Re ≈ 26) we find a free surface shape with sharp indentations which grow with Reynolds number. The position of the sharp indentation shifts slightly downstream while it is getting sharper until the transition Reynolds number (Re ≈ 32) is reached. Here the sharp indentation disappears and the free surface shape changes very abruptly into a smooth sinusoidal one. A further increase of the Reynolds number causes the amplitude of the free surface undulation to diminish continuously.

The inset in the lower right corner shows the flow over the whole channel width from above. The green line in the middle of the inset corresponds to the laser sheet produced by the Ar$^+$-laser described in section 2.4.2. The main flow direction there is from up

---

[2]Video available at http://www.tms.uni-bayreuth.de/videos/surfaces.mp4

to down. At low and high Reynolds numbers, far away from the transition Reynolds number, we find a uniform shape of the free surface over the whole channel width and length. When the transition Reynolds number is approached from low Reynolds numbers particular regions near the side walls appear which seem to differ qualitatively in surface shape from the rest of the film flow in the middle of the channel. The size of these regions grows with increasing Reynolds number until they cover the whole channel, except for the very near wall regions, when the transition Reynolds number is reached. We identify these regions as regions where the surface shape transition from sharp with a strong indentation to a smooth sinusoidal shape has already taken place. We attribute this effect to locally higher flow rates $\dot{q}$ there due to side wall effects. In the very vicinity of the side walls the free surface transition occurs later due to the additional drag coming from the side walls which leads to locally lower flow rates.

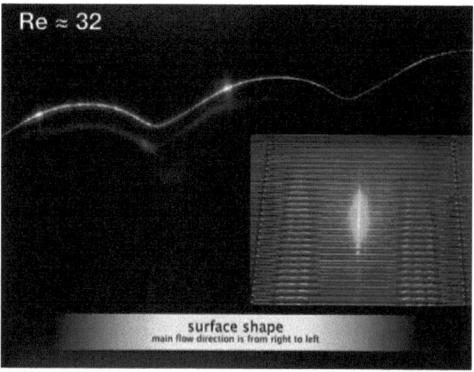

Figure 3.10: Snapshot of a video which illustrates the free surface shape transition at an intermediate Reynolds number. Main flow direction in the main frame is from right to left, in the smaller frame in the lower right corner from up to down. The Reynolds number is increased during the video continuously from Re $\approx$ 26 to Re $\approx$ 34. Channel inclination angle $\alpha = 8\,°$. Video available at URL: http://www.tms.uni-bayreuth.de/videos/surfaces.mp4

We have to note, that the unsteady motion of the free surface, which can be seen in the video, was much weaker but still present under experimental conditions just below the transition Reynolds number. To quantify its impact on the data of the average film thickness and the first two Fourier modes we have carried out a statistical analysis with 1000 images exemplarily for a channel inclination of $8\,°$. Because single images were to grainy to evaluate (see section 2.4.2), each of the 1000 images was created using a running average filter over ten single images. The standard deviation of the calculated Fourier components never exceeded 3.5% which corresponds approximately to the symbol size in Figures 3.6-3.9. Thus, we omit the error bars here. The averaging over 10 single images corresponds to an averaging over 400 ms. This leads to a loss of information regarding very high frequency unsteady motion. Furthermore we observed three–dimensional free surface structures close to the side–walls, as can be seen from video 3.10.

Nevertheless, we state, that the experimental free surface shape data are overall well described by the numerical results, confirming that the assumptions made in section

3.1.1 of a fully developed, purely two–dimensional and steady flow are well met by the experiment. The systematic discrepancies between experimental and numerical data in the average film thickness which can be identified in the Figures 3.6, 3.9(b) and 3.9(c) are apparently due to a misplacement of the ceramic calibration scale perpendicular to the mean flow direction. An effect of this misplacement on higher Fourier components can safely be ruled out. The difference in the plateau of the amplitudes of the second Fourier mode between experiment and numerics is caused by the weak unsteady motion, which is particularly noticeable in the experiment, when the transition Reynolds number is approached from low Reynolds numbers. Because every image had to be averaged over multiple single images the weak jitter of the free surface caused a washing out of sharp indentations which results in an amplitude of the second Fourier component which is systematically too small as can be seen from Figures 3.8 and 3.9.

3.1.3 Physical interpretation and discussion

To characterize this transition we carry out a numerical analysis of the dependence of the local Froude number on the downstream x-coordinate over one bottom period. The Froude number is defined as the ratio of a characteristic velocity to a gravitational wave velocity. In particular, we calculate the local Froude number as $\mathrm{Fr}(x) = u_{\mathrm{loc}}(x)(d(x)g\cos\alpha)^{-1/2}$, where $u_{\mathrm{loc}}(x)$ denotes the local free surface velocity as a function of the downstream coordinate. Figure 3.11 illustrates the local Froude number dependence from the dimensionless downstream coordinate kx for different Reynolds numbers and a fixed channel inclination angle of $\alpha = 8\,^\circ$. The Reynolds number ranges from Re = 5 (lowermost line) to Re = 47.5 (uppermost line).

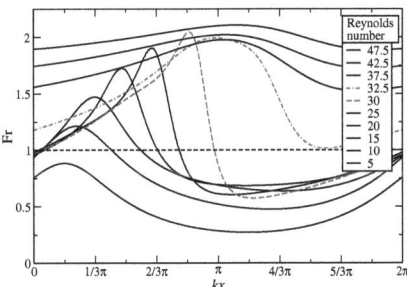

Figure 3.11: Spatial dependence of the local Froude number at different Reynolds numbers. Crests of the topography are at 0 and 2π. The Reynolds number increases continuously from the lowest to the uppermost line. Below transition, the Froude number changes from subcritical to supercritical; beyond transition it remains supercritical. Channel inclination angle $\alpha = 8\,^\circ$. Reprinted with permission from [44]. ©2010, American Institute of Physics.

We find that the Froude number persists globally subcritical (Fr < 1) for low Reynolds numbers below 7 and globally supercritical (Fr > 1) for high Reynolds numbers above 32. In between we observe a jump of the Froude number which undergoes a transition from supercritical to subcritical with increasing the dimensionless downstream coordinate kx. The amplitude of this jump grows with increasing Reynolds number to its maximal size at Re = 30 (red dashed line) until its size decreases rapidly beyond the transition at Re = 32.5 (green dot–dashed line).

The coexistence of sub- and supercritical regions within one flow configuration is typical for hydraulic jumps as reported by Wierschem et al.[41] for a similar flow configuration over an undulated substrate geometry or by Bohr et al.[78, 79] and Bush et al.[80] for a circular hydraulic jump generated by a vertical fluid jet on a horizontal plate. Thus, we conclude, that the flow configuration in the intermediate Reynolds number regime between Re = 7 and Re = 32 can be treated as a hydraulic jump which constantly grows with increasing Reynolds number until it vanishes abruptly at the transition Reynolds number. Evaluation of the local Froude number at other inclination angles did not show any qualitative deviations from the behavior observed for $\alpha = 8\,°$.

The question arises whether the non–monotonous growth and disappearance behavior of the eddy size shown in Figure 3.5 is somehow related to the free surface of the liquid. Figure 3.12 depicts the border positions Re_2 and Re_3 of the eddy free windows from Figure 3.5 for the studied inclinations angles together with the transition Reynolds number. We find that the transition Reynolds number, where the amplitude of the first Fourier component peaks, all other Fourier components change their amplitude discontinuously and the hydraulic jump suddenly disappears, always fits well within the middle of the eddy free window. While this transition Reynolds number shifts to smaller values when the channel inclination angle is increased the eddy free window seems to move with it.

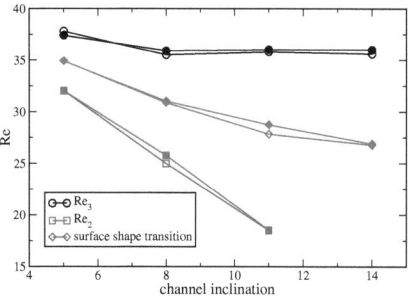

Figure 3.12: Comparison of the position of the free surface shape transition and the eddy–free window position at different channel inclinations. Experimental data are represented by open symbols; numerical data are represented by solid symbols. Reprinted with permission from [44]. ©2010, American Institute of Physics.

The Reynolds number where first eddies appear Re_1 (see Figure 3.5) is independent of the inclination angle α, because the eddy suppressing influence of the free surface shape is not strong enough at such low Reynolds numbers when the channel inclination angle is below $11\,°$. Only for the steepest channel inclination angle investigated, the first Fourier component of the free surface contour is sufficiently strong down to Re = Re_1, what causes the first eddy window between Re_1 and Re_2 to vanish completely.

3.1.4 Conclusions

We have studied the appearance and disappearance of eddies in the troughs of an undulated topography and the free surface shape of the liquid flowing above it. We find that eddies which first appear at rather low Reynolds number can vanish when the Reynolds number is increased. Within an interval, where no eddy is observed the free surface shape undergoes a sharp transition from harmonic to anharmonic and a hydraulic jump which has been established continuously with increasing inertia disappears abruptly.

We have shown, that these eddies are systematically suppressed by the shape of the free surface at rather high Reynolds numbers. Obviously the presence or absence of an hydraulic jump and the amplitude of the higher harmonics in the free surface shape do not play a role for the disappearance of the eddies since they are suppressed below the transition Reynolds number as well as above it. We account the amplitude of the first Fourier component responsible for the suppression of the recirculation areas. When the amplitude of the free surface is large, enough inertia is adjusted for the liquid flowing downhill to penetrate deeply into the troughs of the substrate undulation to break up the eddy structures.

Trifonov[37] also found numerically a disappearance of eddies as illustrated in Figure 3.13 for Reynolds numbers between 130 and 290 in low viscosity flows over a vertically aligned ($\alpha = 90°$) undulated wall. The critical Reynolds Re_c number for the onset of free surface waves in a flow down an inclined channel is given by $\text{Re}_c = (5/6) \cot \alpha$[46, 47]. Inserting $\alpha = 90°$ for films flowing down a vertically aligned wall we find that no linearly stable flow configuration is possible. Therefore, especially flows down a vertically aligned wall at high Reynolds numbers, like the system studied by Trifonov[37], are supposed to be highly unsteady[12]. Bontozoglou[40] has calculated the minimum wall steepness for separation as a function of the Reynolds number for water flowing down a channel inclined by $\alpha = 60°$ with respect to the horizontal. He also found a Reynolds number region at $\text{Re} \sim 200$ at which eddies disappear when the liquid is in resonance with the moderately undulated wall. The groups of Bontozoglou and Aksel showed that an amplification of the free surface amplitude by means of a resonance phenomenon can be achieved at much lower Reynolds numbers[41, 42, 43]. We have chosen the above studied system in a way that the Reynolds numbers at which resonance takes place were expected to become minimal.[42, 43] Thus, we were able to study resonant suppression of eddies in a near steady regime in real experiments.

With this work we directly continue work going on on creating and manipulating of kinematically and/or inertially induced eddies in films.[36] When inertia can be neglected eddies are generated kinematically and sit symmetric in the vallies of the undulation, because the field equations degenerate and become space reversible, unless the boundaries of the liquid do not break the symmetry of the system. The (dimensionless) size of such kinematically induced eddies depends on the geometry parameters like steepness ξ and wavenumber k of the undulation only. Inertially induced eddies are usually asymmetric in film flows and grow monotonously in size when the influence of inertia is increased, be it in film flows over undulated topographies, or for example in bulk flows past a cylinder[77, 81, 82]. Here we show experimentally, as well as numerically, that the eddies can also be diminished in size and even be suppressed completely with increasing Reynolds number under steady, or weakly unsteady flow conditions.

Chapter 3. Two–dimensional film flow

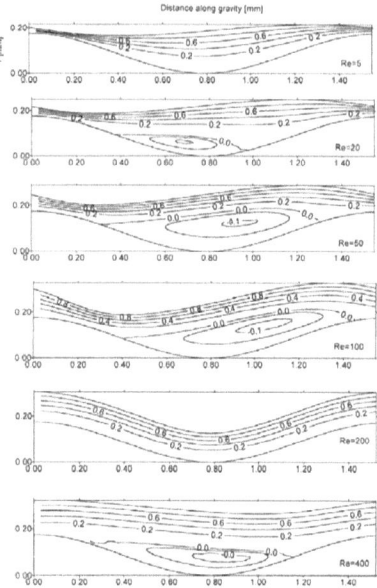

Figure 3.13: Numerically observed streamline patterns by Trifonov[37] for a flow down an undulated, vertically aligned wall ($\alpha = 90°$). At small Reynolds numbers the eddy grows with increasing Reynolds number, then diminishes in size and disappears at Re = 200 before it reappears at even higher Reynolds numbers. Reprinted with permission from [37]. License number 2834701468864. ©1999, Elsevier.

A selective suppression of eddies is of major interest for industrial applications since it can open up new optimized process windows in case when a detachment of the flow undesired. In heat exchanger applications, for example, a detachment of the flow over a corrugated surface would lead to a strong impact on the convective rate of heat transport. In environmental systems, particles within a recirculation area are cut off from the rest of the flow and thus also from potentially necessary nutrient substances. Different from applying external forces to the flow, generating a strong fundamental harmonic of the free surface just by exciting resonance seems to be a good framework in general, to suppress eddies in gravity–driven film flows over undulated topography.

Chapter 4

Three–dimensional film flow

4.1 Basic flow

4.1.1 Governing equations

We consider a steady gravity–driven film flow down a flat channel of finite width B which is inclined by an angle α with respect to the horizontal as illustrated in Figure 4.1. The geometry of the flow configuration is defined by the channel width B, the contact angle between the liquid and the channel side wall θ, the film thickness in the middle of the channel H and the resulting position of the free surface $h(y)$. The capillary elevation height just at the side wall is denoted by Δh. For infinite long channels the in- and outflow effects are negligible and the flow can be assumed to be unidirectional $\mathbf{u} = u(y,z)\hat{e}_x$. The orientation of the (x,y,z)–coordinate system is defined by the unit vectors $\hat{e}_x$, $\hat{e}_y$ and $\hat{e}_z$ and its point of origin lies just on the surface of the substrate in the middle of the channel as illustrated in Figure 4.1.

Inserting this assumption into the steady Navier–Stokes equations and the continuity equation

$$(\mathbf{u} \cdot \nabla)\mathbf{u} = -\frac{1}{\rho}\nabla p + \mathbf{g} + \nu \nabla^2 \mathbf{u}, \qquad \nabla \cdot \mathbf{u} = 0, \tag{4.1}$$

leads to

$$0 = \eta\left(\frac{\partial^2 u}{\partial y^2} + \frac{\partial^2 u}{\partial z^2}\right) + \rho g \sin\alpha, \tag{4.2}$$

$$0 = -\frac{\partial p}{\partial y}, \tag{4.3}$$

$$0 = -\frac{\partial p}{\partial z} - \rho g \cos\alpha \tag{4.4}$$

and a continuity equation which fulfilled in a trivial way by the kinematic assumption. The non–linear terms on the left hand side in the steady Navier–Stokes equations (4.1) disappear because the unidirectional flow field u does depend on the y- and on the z-coordinate only.

The no–slip boundary conditions at the bottom ($z = 0$) and the side walls ($y = \pm B/2$)

Chapter 4. Three–dimensional film flow

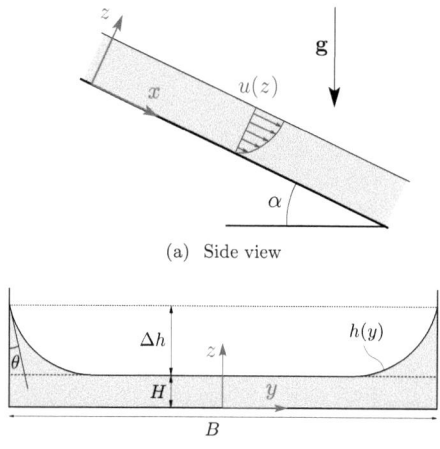

(a) Side view

(b) Cross–sectional view

Figure 4.1: Channel geometry illustrating side wall effects on the flow and the orientation and position of the (x, y, z)–coordinate system. Reprinted with permission from [66]. ©2011, American Institute of Physics.

read
$$u(y, z = 0) = 0, \qquad (4.5)$$
$$u(y = \pm B/2, z) = 0. \qquad (4.6)$$

The no penetration condition at the rigid walls is trivially fulfilled because the flow has been postulated to be unidirectional. Additionally to the boundary conditions at the walls a kinematic boundary condition at the free surface

$$\left. \mathbf{n} \cdot \mathbf{u} \right|_{z=h(y)} = 0 \qquad (4.7)$$

basically demands the free surface contour to be a streamline, or in other words liquid particles must not leave the free surface. When the viscosity of air is neglected the dynamic boundary condition, which takes care of the balance of stresses at the free surface, reads

$$\left[(p - p_0) - \frac{\sigma}{R} \right] \mathbf{n} = \underline{\mathbf{T}} \cdot \mathbf{n} \Big|_{z=h(y)}, \qquad (4.8)$$

with the stress tensor
$$\underline{\mathbf{T}} = T_{ij} = \eta \left(\frac{\partial u_i}{\partial x_j} + \frac{\partial u_j}{\partial x_i} \right) \qquad (4.9)$$

and the outer unit normal vector
$$\mathbf{n} = \frac{\nabla h}{\|\nabla h\|} = \frac{\hat{e}_z - \frac{\partial h}{\partial y} \hat{e}_y}{\sqrt{1 + \left(\frac{\partial h}{\partial y} \right)^2}}. \qquad (4.10)$$

47

The pressure of the surrounding air is denoted by p_0. The curvature of the free surface κ which is the inverse radius of curvature of the free surface R can be calculated to

$$\kappa = \frac{1}{R} = \nabla \cdot \mathbf{n} = -\frac{d}{dy}\frac{\frac{\partial h}{\partial y}}{\sqrt{1+\left(\frac{\partial h}{\partial y}\right)^2}} = -\frac{\frac{\partial^2 h}{\partial y^2}}{\left[1+\left(\frac{\partial h}{\partial y}\right)^2\right]^{3/2}}. \tag{4.11}$$

Inserting (4.10) and (4.11) into the dynamic boundary condition (4.8) leads to two components of the dynamic boundary condition normal and tangential to the free surface

$$(p - p_0)\Big|_{z=h(y)} = -\sigma \frac{d}{dy}\frac{\frac{\partial h}{\partial y}}{\sqrt{1+\left(\frac{\partial h}{\partial y}\right)^2}}, \tag{4.12}$$

$$\frac{\partial u}{\partial z}\Big|_{z=h(y)} = \frac{\partial u}{\partial y}\frac{\partial h}{\partial y}. \tag{4.13}$$

Dimensionless formulation

To reformulate the problem in a dimensionless form we have to find some reference quantities for scaling. We take the free surface velocity

$$u_r = \frac{\rho g \sin \alpha H^2}{2\eta} \tag{4.14}$$

from the well known Nusselt solution [73] as a reference for all velocities. Hydrostatic pressure is taken as a reference for the pressure

$$p_r = \rho g \cos \alpha. \tag{4.15}$$

We introduce a generalized capillary length L

$$L = \sqrt{\frac{2\sigma}{\rho g \cos \alpha}}, \tag{4.16}$$

which takes care of a reduced gravitational acceleration perpendicular to the channel due to its inclination α and serves as a reference for all lengths to resolve effects within the capillary elevation. In the following all quantities which are labeled by a $\tilde{\ }$ denote dimensionless variables which are scaled with the above reference quantities. Furthermore, we define a dimensionless capillary range l

$$l = \frac{L}{B/2} = \frac{2L}{B}. \tag{4.17}$$

Applying the above scalings to the Navier–Stokes equations and the boundary conditions lead to a dimensionless formulation of the system in the following form. The Navier–Stokes equations read:

$$0 = \frac{\partial^2 \tilde{u}}{\partial \tilde{y}^2} + \frac{\partial^2 \tilde{u}}{\partial \tilde{z}^2} + 2, \tag{4.18}$$

$$0 = -\frac{\partial \tilde{p}}{\partial \tilde{y}}, \tag{4.19}$$

$$0 = -\frac{\partial \tilde{p}}{\partial \tilde{z}} - 1. \tag{4.20}$$

Chapter 4. Three–dimensional film flow

The no–slip boundary conditions at the bottom and the side walls changes to

$$\tilde{u}(\tilde{y}, \tilde{z} = 0) = 0, \qquad (4.21)$$
$$\tilde{u}(\tilde{y} = \pm 1/l, \tilde{z}) = 0 \qquad (4.22)$$

and the tangential and normal components of the dynamic boundary condition now read

$$\left.\frac{\partial \tilde{u}}{\partial \tilde{z}}\right|_{\tilde{z}=\tilde{h}(\tilde{y})} = \frac{\partial \tilde{u}}{\partial \tilde{y}} \frac{d\tilde{h}}{d\tilde{y}}, \qquad (4.23)$$

$$\left.(\tilde{p} - \tilde{p}_0)\right|_{\tilde{z}=\tilde{h}(\tilde{y})} = -\frac{1}{2} \frac{d}{d\tilde{y}} \frac{\frac{\partial \tilde{h}}{\partial \tilde{y}}}{\sqrt{1 + \left(\frac{\partial \tilde{h}}{\partial \tilde{y}}\right)^2}}. \qquad (4.24)$$

Free surface shape

We introduce an additional boundary condition taking care of the contact angle θ between the liquid and the side wall

$$\left.\frac{\partial \tilde{h}}{\partial \tilde{y}}\right|_{\tilde{y}=\pm 1/l} = \pm \cot \theta \qquad (4.25)$$

and a decomposition of the free surface shape $\tilde{h}(\tilde{y})$ into a constant part which is equal to the film height in the middle of the channel $\tilde{H}$ and a part ζ depending on $\tilde{y}$ which takes care of the capillary elevation in the vicinity of the side wall

$$\tilde{h}(\tilde{y}) = \tilde{H} + \zeta(\tilde{y}). \qquad (4.26)$$

Evaluating ζ at $\tilde{y} = \pm 1/l$ by inserting the boundary condition (4.25) into (4.24) leads to the capillary elevation height depicted in Figure 4.1(b) [66]

$$\Delta \tilde{h} = \zeta(\tilde{y} = \pm 1/l) = \sqrt{1 - \sin \theta}. \qquad (4.27)$$

The free surface shape can be obtained by integration of equation (4.24) as described in detail by Scholle and Aksel[63] or Haas et al.[66]. With the abbreviation

$$G(x) := x - \frac{1}{2\sqrt{2}} \ln\left(\frac{\sqrt{2}+x}{\sqrt{2}-x}\right), \qquad (4.28)$$

the function for the film elevation $\zeta(\tilde{y})$ can be written down in an implicit form

$$\tilde{y}(\zeta) = \tilde{h}^{-1}(\zeta) = \begin{cases} -1/l + \left[G\left(\sqrt{1+\sin\theta}\right) - G\left(\sqrt{2-\zeta^2}\right)\right] & \tilde{y} \in [-1/l, 0], \\ 1/l - \left[G\left(\sqrt{1+\sin\theta}\right) - G\left(\sqrt{2-\zeta^2}\right)\right] & \tilde{y} \in [0, 1/l]. \end{cases} \qquad (4.29)$$

Velocity field

The velocity field can be described by the following ansatz which is a solution of equation (4.18) and already fulfills the no-slip boundary conditions at the side walls

$$\tilde{u} = \frac{4}{l^2} \sum_{n \in \mathbb{N}^+} \left[D_n e^{l c_n \tilde{z}} + E_n e^{-l c_n \tilde{z}} - \frac{(-1)^n}{c_n^3}\right] \cos(l c_n \tilde{y}), \qquad (4.30)$$

4.1. Basic flow

$$c_n = \left(n - \frac{1}{2}\right)\pi. \tag{4.31}$$

The additional no-slip condition at the substrate (4.21) leads to n equations for the vector elements D_n and E_n

$$D_n + E_n = \frac{(-1)^n}{c_n^3}. \tag{4.32}$$

Inserting the ansatz for the velocity field (4.30) into the tangential part of the dynamic boundary condition (4.23) leads to an infinite system of algebraic equations for the coefficients D_n

$$\sum_{n \in \mathbb{N}^+} G_{nm} D_n = d_m, \tag{4.33}$$

with the matrix elements

$$G_{nm} = \int_{-1/l}^{1/l} 2\cosh(lc_n \tilde{h}(\tilde{y})) \sin(lc_n \tilde{y}) \sin(lc_m \tilde{y}) \mathrm{d}\tilde{y} \tag{4.34}$$

and the vector elements

$$d_m = \sum_{n \in \mathbb{N}^+} \int_{-1/l}^{1/l} \left[\frac{(-1)^n}{c_n^3} \left(e^{-lc_n \tilde{h}(\tilde{y})} + lc_n \tilde{h}(\tilde{y}) \right) \right] \sin(lc_n \tilde{y}) \sin(lc_m \tilde{y}) \mathrm{d}\tilde{y}$$

$$- \frac{l}{c_m} \sum_{n \in \mathbb{N}^+} \int_{-1/l}^{1/l} \frac{(-1)^n \tilde{h}(\tilde{y})}{c_n} \left[\cos(lc_n \tilde{y}) \cos(lc_m \tilde{y}) \right] \mathrm{d}\tilde{y}. \tag{4.35}$$

The coefficients D_n and E_n have been calculated using MATLAB® [83] truncating the infinite system of algebraic equations to a certain order $N \in \mathbb{N}^+$. The accuracy of the power series expansion has been assured by the demand $D_n, E_n < 10^{-6}$.[66]

The theoretical derivation and the implementation of the procedure of solution described above has been done and all theoretical results presented in this chapter have been calculated by André Haas[66].

4.1.2 Flow type classification

Depending on the magnitude of the capillary elevation compared to the film height H and the channel width B it is useful to distinguish different flow types as illustrated in Figure 4.2. For a channel of infinite extent the flow configuration is equivalent the two–dimensional case and the solution of the velocity field equals the well known Nusselt solution. In this case we define the Reynolds number as

$$\mathrm{Re}_{(a)} = \mathrm{Re}_{2D} = \frac{u_s H}{\nu}. \tag{4.36}$$

Because the free surface velocity of a Nusselt film flow can easily be calculated to

$$u_s = \frac{g \sin \alpha H^2}{2\nu}, \tag{4.37}$$

the Reynolds number can also be expressed as a function of the film height H

$$\mathrm{Re}_{2D} = \frac{g \sin \alpha H^3}{2\nu^2}. \tag{4.38}$$

Chapter 4. Three–dimensional film flow

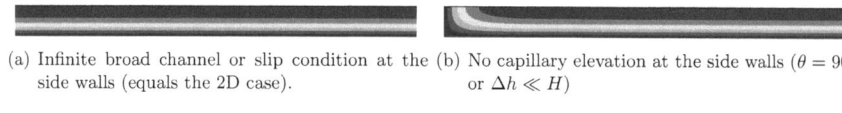

(a) Infinite broad channel or slip condition at the side walls (equals the 2D case).

(b) No capillary elevation at the side walls ($\theta = 90°$ or $\Delta h \ll H$)

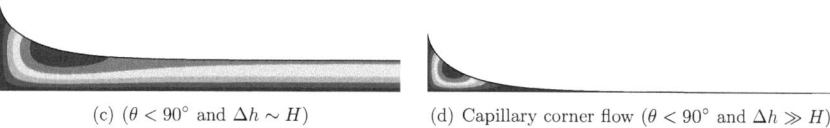

(c) ($\theta < 90°$ and $\Delta h \sim H$)

(d) Capillary corner flow ($\theta < 90°$ and $\Delta h \gg H$)

Figure 4.2: Cross sectional velocity profiles of different channel flow types. The velocity is color coded: blue corresponds to slow and red corresponds to fast. Reprinted with permission from [66]. ©2011, American Institute of Physics.

As soon as side walls are introduced the liquid in the vicinity of the side wall gets decelerated due to the additional no–slip condition at the boundary. Now, the free surface velocity is not independent of the crosswise coordinate y anymore and it is useful to define the Reynolds number in terms of the corresponding volume flux $\dot{V}$

$$\text{Re}_{(b)} = \frac{3}{2B\nu} \int_0^{f(y)} \int_{-B/2}^{B/2} u(y,z) \mathrm{d}y \mathrm{d}z = \frac{3\dot{V}}{2B\nu} < \text{Re}_{2D}. \qquad (4.39)$$

Compared to the scenario depicted in Figure 4.2(a) less liquid is transported due to the additional drag at the side walls. Therefore, the Reynolds number of a film in this case is alway smaller than the Reynolds number of a two–dimensional film of the same film thickness H.

When H is decreased, the contact angle θ between the liquid and the side wall becomes an important factor. The influence of capillarity leads to a capillary elevation of the liquid, when θ is smaller than $90°$. Due to this locally thicker film a velocity overshoot close to the side walls may show up when the additional film thickness wins over the additional drag coming from the no–slip condition at the side walls (see Figure 4.2(c)). In the limit of vanishingly thin films ($H \to 0$) the flow degenerates to a capillary corner flow as depicted in Figure 4.2(d). Now most of the liquid is transported close to the side walls in the capillary elevation and the Reynolds number is obviously larger than in the two–dimensional case

$$\text{Re}_{(d)} = \frac{3\dot{V}}{2B\nu} > \text{Re}_{2D} \to 0, \quad H \to 0. \qquad (4.40)$$

When the contact angle between the liquid and the side wall is larger than $90°$ the capillary elevation Δh becomes negative and no velocity overshoot can be observed. Therefore, we restrict our studies without loss of generality to liquids with wetting properties only.

4.1.3 Flow rate study

Since the presence of side walls and the resulting capillary elevation has a significant impact on the flow rate, a study on the important parameters is to be carried out. The volume flux of a two–dimensional case $\dot{V}_{2D}$ which is illustrated in Figure 4.2(a) is taken as a reference.

Figure 4.3 illustrates the inequality (4.39) for two different capillary ranges, one matching the experimental setup ($l = 0.028$) and one referring to the capillary range of a narrower (and/or steeper) channel ($l = 0.1$). When no capillary elevation is present, as depicted in Figure 4.2(b) the additional drag at the side walls leads to an overall decrease of the transported liquid depending on the dimensionless film height $\tilde{H}$ and the capillary range l. The amount of missing volume flux increases with decreasing channel width B and increasing dimensionless film height $\tilde{H}$. Both an increasing dimensionless film height and a narrowing of the channel lead to an increase of the relative part of the side wall area which results in a stronger impact of the side wall presence on the volume flux. Therefore, especially for narrow channels and film thicknesses which are large compared to the capillary length L the influence of the side walls on the volume flux cannot be neglected.

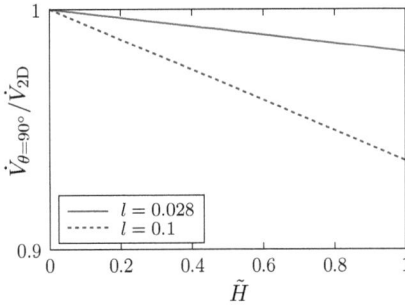

Figure 4.3: Decrease of the volume flux due to the additional no–slip condition at the side walls (without capillarity), as it is depicted in Figure 4.2(b), compared to the two–dimensional case, which is depicted in Figure 4.2(a). Reprinted with permission from [66]. ©2011, American Institute of Physics.

In the case depicted in Figure 4.2(c) it is not possible to make a similar general statement on the Reynolds number or the volume flux like in the equations (4.39) or (4.40), because both effects of one an additional drag coming from the side wall and two a velocity overshoot have a competing influence of the same order on the flow rate as illustrated in Figure 4.4.

Figure 4.5 compares the amount of transported liquid when a capillary elevation due to the presence of side walls is considered with the two–dimensional case. The contact angles θ are chosen to fit the experimental setup as described in section 2.2. For small film thickness $\tilde{H}$ the influence of the capillary elevation and the resulting velocity overshoot becomes the most important transport mechanism. The ratio of $\dot{V}/\dot{V}_{2D}$ becomes larger than one and even diverges for $\tilde{H} \to 0$ because $\dot{V}_{2D}$ then also tends to zero. For constant film height $\tilde{H}$ the presence of a capillary elevation becomes obviously more important for narrower channels or larger capillary ranges l (see Figure 4.5(a)). Remember, that the two–dimensional case is equal to the case $l \to 0$. The influence of the capillary range on the volume flux increases with decreasing film height. One finds an explicit transition film height denoted by $\tilde{h}_t$, which is independent of the capillary length, where the influences of the velocity overshoot due to capillary elevation and the influence of the no–slip condition

Chapter 4. Three–dimensional film flow

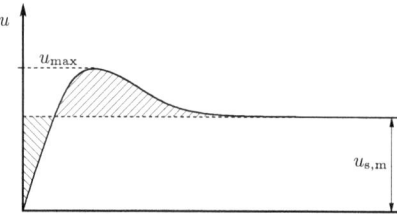

Figure 4.4: Free surface velocity profile in flow direction showing velocity overshoot and defect compared to the plane flow with the same film height. Reprinted with permission from [70]. ©2011, American Institute of Physics.[70]

at the wall on the normalized volume flux $\dot{V}/\dot{V}_{2D}$ just cancel each other.

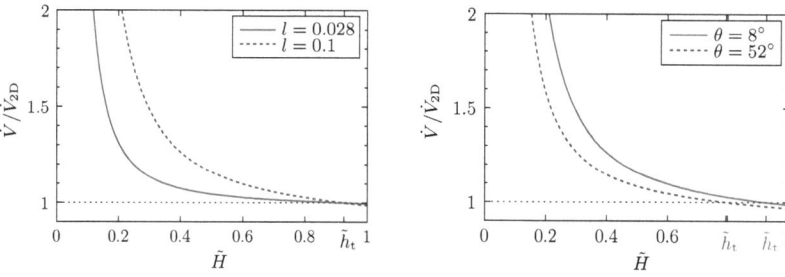

(a) Variation of the capillary range l at fixed contact angle $\theta = 8°$.

(b) Variation of the contact angle θ at fixed capillary range $l = 0.1$.

Figure 4.5: Influence of capillary effects at the side walls on the normalized volume flux. Reprinted with permission from [66]. ©2011, American Institute of Physics.

The impact of the contact angle θ on the volume flux is illustrated in Figure 4.5(b). Similar to the capillary range in Figure 4.5(a) also the influence of the contact angle on the volume flux is small for film heights $\tilde{H}$ of the order of one but gains in importance the thinner the liquid film gets. One finds that smaller contact angles lead to a larger velocity overshoot and therefore to a larger volume flux. The transition film height $\tilde{h}_t$ is not independent of the contact angle.

For large $\tilde{H}$ the ratio of $\dot{V}/\dot{V}_{2D}$ becomes smaller than one because the influence of capillarity on the velocity field looses importance. The volume flux $\dot{V}$ tends to $\dot{V}_{\theta=90°}$ for $\tilde{H} \to \infty$ which is always smaller than $\dot{V}_{2D}$ as depicted in Figure 4.3.

Figure 4.6 shows the dependence of the transition film height $\tilde{h}_t$ on the contact angle θ. For a contact angle of $\theta = 90°$ no capillary elevation is present and thus no velocity overshoot can be observed. Equality of the volume flux of the 2D case $\dot{V}_{2D}$ and the volume flux $\dot{V}$ can only be reached, when also the drag influence of the side wall tends to zero which is only the case in the limit $\tilde{h} \to 0$. Decreasing the contact angle leads to a monotonous increase of the transition film thickness to finite values below one. Under perfect wetting conditions ($\theta = 0°$) the transition film height reaches a value of approximately $\tilde{h}_t \approx 0.92$.

To summarize: Treating a channel of finite width as two–dimensional always leads to

4.1. Basic flow

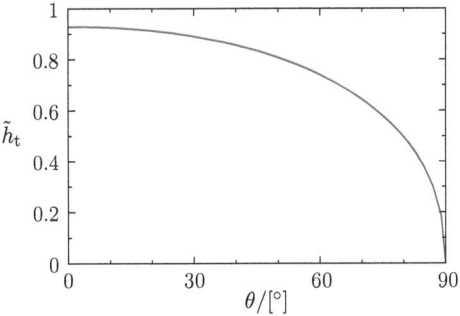

Figure 4.6: Dependence of the transition film thickness $\tilde{h}_t$ on the contact angle θ. Reprinted with permission from [66]. ©2011, American Institute of Physics.

an overestimation of the volume flux (or the Reynolds number) when the film thickness H is larger than the generalized capillary length L (or above the red curve in Figure 4.6) due to the additional drag coming from the side walls. This discrepancy becomes large especially for narrow channels. When the film thickness is small or of the same order as the generalized capillary length the presence of a capillary elevation gains in importance one has to take the contact angle θ into account. The effect of a resulting velocity overshoot competes with the additional drag at the side wall. For thin films and small contact angles (below the red curve in Figure 4.6) treating the film as two–dimensional leads to an underestimation of the volume flux.

4.1.4 Velocity field

Figure 4.7 shows a comparison of the theoretical and measured velocity profiles for three films of different heights H and two different static contact angles θ. As liquid *Elbesil* silicone oil B1000 which is described in section 2.1 was used. For each flow configuration the side wall distance d_s dependence of the velocity profiles was measured by a Laser Doppler Velocimeter described in section 2.4.4 at three different measurement heights H_m.

The error bars of the measured data denote the root mean square error of the mean value of all detected velocity signals in each measurement volume. Because the number of evaluable counts per time decreases with the speed of the liquid in the measurement volume, the measurement time has been adopted to the flow velocity to get reasonable signal to noise ratios especially in near wall regions. Additionally, more points have been recorded in the vicinity of the wall to resolve the velocity overshoot.

The overall agreement between the measured data and the calculated values for the flow velocities is excellent. The small deviations which never exceed the root mean square error bars are in the most cases of statistically nature. Systematic discrepancies such as in Figure 4.7(a) can be explained by errors in determining the distance between the channel bottom and the measurement volume H_m or in determining the film height in the middle of the channel H.

For large side wall distances the measured and calculated velocity profile corresponds

Chapter 4. Three–dimensional film flow

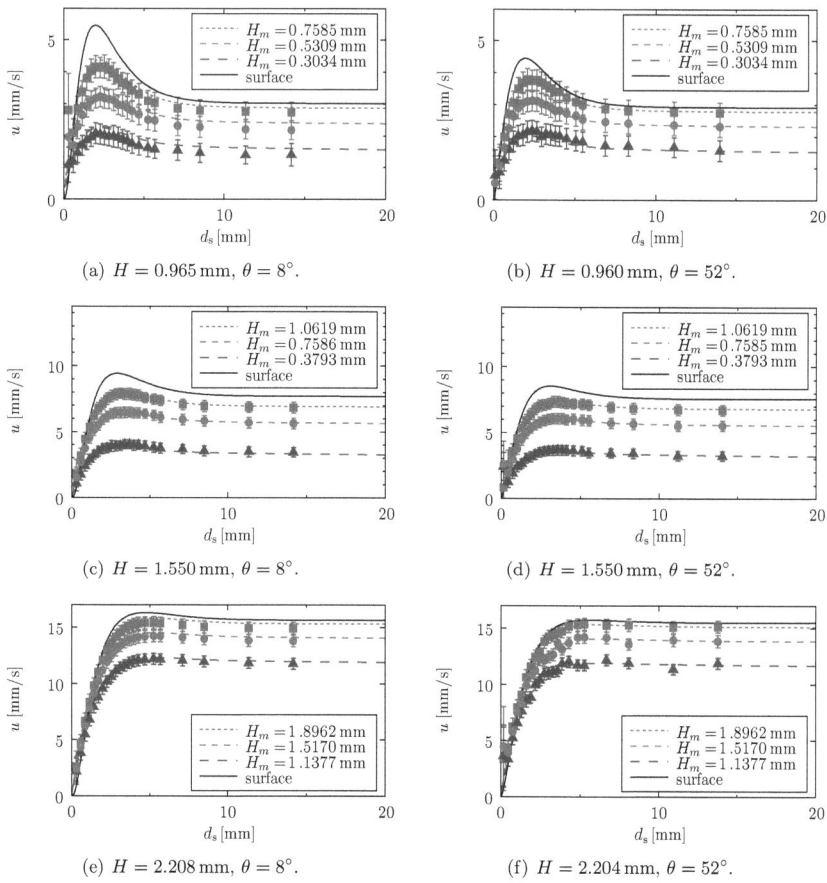

Figure 4.7: Comparison of measured (points) and calculated (lines) velocity profiles for different film heights H, measurement heights H_m and contact angles θ. Reprinted with permission from [66]. ©2011, American Institute of Physics.

to the Nusselt solution. In the near wall region a velocity overshoot is observed whose magnitude, quantified by the ratio of the highest velocity u_{max} and the surface velocity in the middle of the channel $u_{\text{s,m}}$, depends strongly on the film thickness H and the contact angle θ.

Figure 4.8(a) shows the dependence of the magnitude of the velocity overshoot on the film thickness $\tilde{H}$ for both contact angles θ measured. For thick films the ratio of $u_{\text{max}}/u_{\text{s,m}}$ tends to one, meaning that no velocity overshoot can be observed. When the film thickness $\tilde{H}$ is approximately 0.5 or less the magnitude of the velocity overshoot becomes considerable and depends on the contact angle θ as also shown in Figure 4.7. As the film thickness is decreased further the velocity overshoot diverges, because the free surface velocity in the middle of the channel tends to zero. The flow then degenerates to a capillary corner flow as illustrated in Figure 4.2(d).

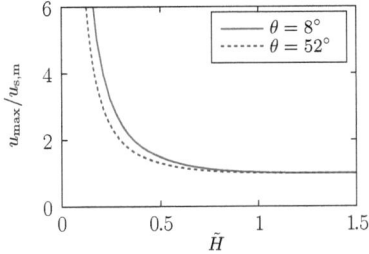
(a) Normalized maximal free surface velocity.

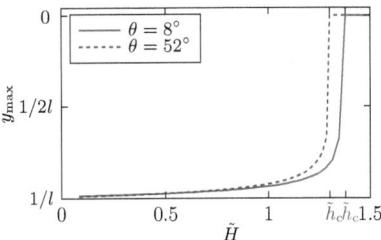
(b) Position of the maximum free surface velocity.

Figure 4.8: Influence of the film thickness on the velocity overshoot. Reprinted with permission from [66]. ©2011, American Institute of Physics.

Not only the magnitude of the velocity overshoot but also its location depends on the film height $\tilde{h}$ as well as on the contact angle θ. Figure 4.8(b) shows film height dependence of the $\tilde{y}$-position $\tilde{y}_{\max}$ where the largest free surface velocity $u_{\max}$ is located. An increase of the film height leads to a shift of the maximum free surface velocity towards the middle of the channel until the velocity overshoot disappears and the maximal velocity is located in the middle of the channel at $\tilde{y} = 0$. The critical film thickness at which the velocity overshoot disappears, labeled in Figure 4.8(b) by $\tilde{h}_{\mathrm{c}}$, shifts with increasing contact angle θ to smaller values.

Both the position and the magnitude of the velocity overshoot depend on the contact angle θ and the film height H. Therefore, we introduce a combined dimensionless parameter r which can be attributed to a certain shape of the free surface velocity profile and depends on the dimensionless capillary elevation height $\Delta\tilde{h}(\theta)$ and the dimensionless film height $\tilde{H}$ in the form of [66]

$$r = \frac{\Delta\tilde{h}(\theta)^{c_1}}{\tilde{H}^{c_2}}. \tag{4.41}$$

The free parameters c_1 and c_2 were obtained by fitting the results of additional simulations: $c_1 = 0.0435 \pm 0.002$ and $c_2 = 0.9814 \pm 0.0176$.[66] With these parameters inserted in equation (4.41) all experimental setups with the same parameter r show the same behavior for the velocity overshoot. Thus, we can now find a critical ratio r_c for the onset of a velocity overshoot to $r_c = 0.733 \pm 0.002$. Inserting r_c into the equation (4.27) for the capillary elevation height leads to an empirical threshold for the onset of a velocity overshoot in terms of a critical film thickness

$$\tilde{h}_c = r_c^{-1/c_2}(1 - \sin\theta)^{c_1/(2c_2)} \tag{4.42}$$

which is illustrated in Figure 4.9.

The critical film thickness $\tilde{h}_c$ is larger than transition film thickness $\tilde{h}_t$ (compare Figures 4.6 and 4.9), because $\tilde{h}_c$ describes the film thickness where a velocity overshoot just emerges. The magnitude of the velocity will not become sufficiently strong to balance the no–slip condition at the until the film height is decreased further to the transition film height $\tilde{h}_t$ which, therefore, has always to be smaller than $\tilde{h}_c$.

Chapter 4. Three–dimensional film flow

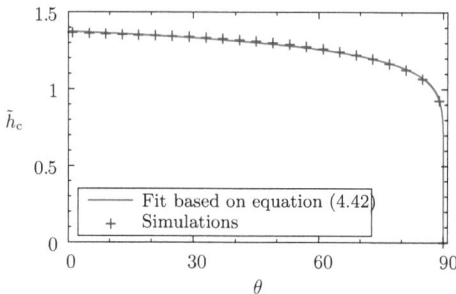

Figure 4.9: Dependence of the critical film thickness $\tilde{h}_c$ on the contact angle θ. Reprinted with permission from [66]. ©2011, American Institute of Physics.

4.1.5 Free surface shape

Steady flow

The steady solution for the free surface shape $\tilde{h}$ of a liquid flowing down an inclined channel with side walls and a contact angle θ there given in an implicit form by equation (4.29) is illustrated in Figure 4.10.

The experimental setup for the correspondent measurement technique is described in section 2.4.2. To get a uniformly bright light sheet from the fluorescent particles in the liquid we have superimposed five single images in the case of steady flow.

Since we were not able to determine the position of the side wall with the desired accuracy from the images, because only fluorescent light was detected by the camera, a single paramater fit with freedom in the y-direction was performed to match the experimental data set with the theory. We have restricted our measurements to the near wall region $d_s \lesssim 3L$ to increase the resolution in the capillary elevation. For side wall distances $d_s > 3L$ the surface shape of all contact angles, including the flat case of $\theta = 90°$, basically coincides. For both measured contact angles θ the calculated and detected free surface shape shows perfect agreement.

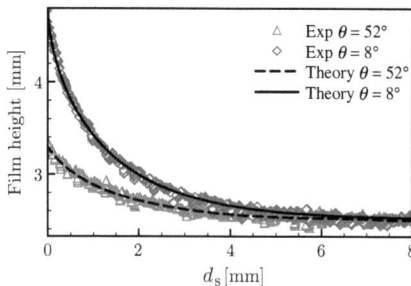

Figure 4.10: Theoretical and experimental shape of the free surface in the vicinity of the side wall. Reprinted with permission from [66]. ©2011, American Institute of Physics.

4.1. Basic flow

Draining flow

Next, the question whether equation (4.29) is capable of describing flows with a slowly decreasing film height is addressed. Here, we want to investigate the case of a draining flow, which is of major interest in nature and engineering processes. We simulate the draining flow process by switching off the pump. The initial film height H in the middle of the channel was 2.5 mm. During the draining the free surface shape of the liquid was recorded in the vicinity of the side wall with a capturing rate of 8 Hz. The position of the free surface was determined from the upper edge of the bright sheet in each averaged image obtained from a moving average of 5 single images.

Figure 4.11 shows a comparison of the free surface shape of the steady flow before switching off the pump and for the draining flow 13.3 s after switching off the pump. The data sets have been shifted vertically to coincide at the side wall distance $d_s = 0$ position. It directly becomes obvious, that the free surface shape of a draining flow is not identical with the steady flow case. During the draining flow one observes for both contact angles an increase of the capillary elevation height Δh. The change of Δh from the steady case to the draining case is more significant for the system with the larger static contact angle. This suggests to describe the shape of the dynamic case by introducing a new free parameter the dynamic or receding contact angle $\theta_d < \theta$ which replaces θ in equation (4.25).

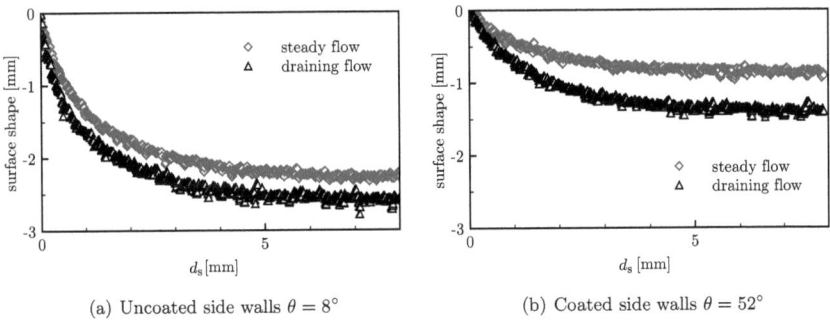

(a) Uncoated side walls $\theta = 8°$ (b) Coated side walls $\theta = 52°$

Figure 4.11: Comparison of the steady free surface shape with the free surface shape of the draining flow 13.3 s after switching off the pump. Reprinted with permission from [66]. ©2011, American Institute of Physics.

Figure 4.12 shows the evolution of film height in the middle of the channel and in the vicinity of the side walls. In the middle of the channel only one data set is plotted for both contact angle systems, because it did not show any significant contact angle dependence. The free-surface position in the near–wall region is more noisy for two reasons. One, the fact, that more particles are passing the laser sheet per time interval in the middle of the channel than in the very near–wall region because the particles are faster there, which results in a better ensemble averaging. Two, the spatial averaging in the middle of the channel has ben done with 150 pixels and at the side walls averaging was done with 15 pixels only, because the film height $h(y)$ does not show a strong y-dependence in the middle of the channel in contrast to the near wall region.

Chapter 4. Three–dimensional film flow

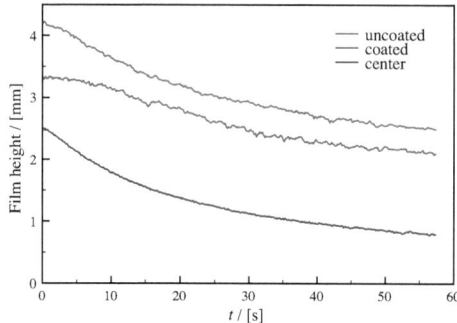

Figure 4.12: Evolution of the free surface position in the middle of the channel (blue) and of the near wall region (red and green) after switching off the pump.

From Figure 4.12 we can find, that the film height in the middle of the channel drops faster at the beginning of the draining process than in the near wall region. Especially, the system with coated side-walls shows a distinctive lag in time. Therefore, the capillary elevation height starts to increase and a new dynamic contact angle θ_d is build up.

However, even if a variation of the contact angle is allowed, equation (4.29) is still not capable to describe the experimental data of the draining flow case as quasi steady states, even though the settling speed is some orders of magnitudes smaller than the flow velocities in x-direction. Also the corresponding Reynolds number for the settling defined as $\text{Re}_s = wH/\nu$, where w is the velocity in z-direction, is only of the order of 10^{-4}. Hence we conclude, that in addition to the local effect of a dynamic contact angle there must exist another (global) effect which is responsible for the free surface shape deformation.[66]

To identify the reason for this dynamic free surface shape deformation, additional numerical simulations, modeling the experimental setup of a draining channel flow, of the time dependent three–dimensional Navier–Stokes equations have been carried out by André Haas as described detailed in [66] with the open source CFD (computational fluid dynamics) code OpenFOAM[84]. The contact angles have been kept constant at $\theta = 8°$ and $\theta = 52°$. Figure 4.13 shows the free surface shapes at four subsequent time steps. To aid comparison, the curves have been shifted in z–direction, in a way, that the triple point liquid/air/side wall coincides with the steady case ($t = 0$). Since the major interest lies on the free surface shape in the vicinity of the side wall, and to aid comparison with Figure 4.11, subviews in Figure 4.13 show the free surface shapes in the near wall region, only.

In contrast to the experiments the capillary elevation Δh diminishes with time, because no dynamic contact angle $\theta_d \leq \theta$ at the side wall has been taken into account in the numerical simulations. In the vicinity of the side wall an indentation is formed, which becomes more and more pronounced, as time advances. This indentation is a result of the velocity overshoot in the capillary elevation, which causes a faster drain there.

This result is similar to the findings of Aksel[18] who found experimentally an indentation of the liquid's free surface at the side walls near to the outflow edge of an inclined

4.1. Basic flow

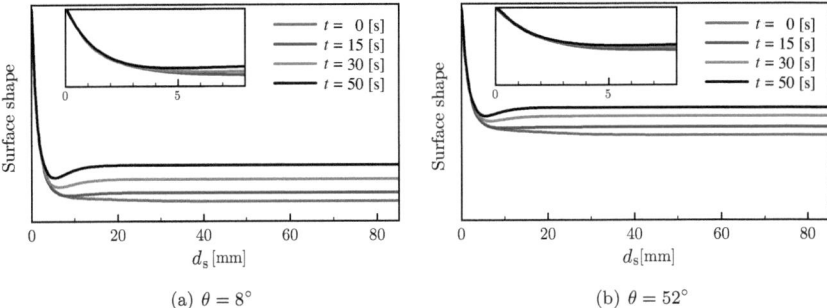

Figure 4.13: Numerical simulations of the free surface shape of a draining flow over one half of the channel $0 \leq d_s \leq B/2$. The inlaid subviews to show the local impact on the free surface shape in the proximity of the side walls. [66]

channel. In both cases, at an outflow edge and in the draining case, one finds an interplay between two different radii of curvature one in the yz–plane and one in the xz–plane which causes this complex free surface shape.

Based on experimental and numerical results two different effects on the free surface shape coming from the side walls have been identified, which are not included in equation (4.29). Due to the faster drain near the side walls an indentation of the surface shape develops, which grows as time advances, because the liquid film becomes thinner, and the velocity overshoot becomes more and more pronounced (See Figure 4.8(a)). Additionally, a dynamic contact angle is formed, which is smaller than the static contact angle. Thus, the capillary elevation in the draining case is larger, than the capillary elevation under steady flow conditions.

4.1.6 Conclusions

The influence of side walls and the contact angle of the liquid there on the velocity profile and flow rates has been studied theoretically as well as experimentally. We find that the influence of side walls is twofold. On the one hand, the additional no–slip condition at the side wall leads to additional drag. On the other hand, capillarity leads, under wetting conditions, to an elevation of the liquid and thus to a locally thicker film. Depending on the contact angle between the liquid and the side wall θ and the dimensionless film thickness $\tilde{H}$ this elevation may lead to a velocity overshoot in the vicinity of the side walls.

Both effects also contribute oppositional to the overall flow rate. We find, that treating a channel flow bounded by side walls as two–dimensional might lead to an under– as well as to an overestimation of the flow rate. For vanishingly thin films obviously most liquid is transported close to the side walls of the channel, which means, that neglecting the three–dimensionality of the flow would lead to an underestimation of the flow rate. When the film thickness is large compared to the capillary elevation the drag influence of the side walls wins over the influence of the capillary elevation on the flow rate and neglecting the influence of side walls would lead to an overestimation of the flow rate. We have

Chapter 4. Three-dimensional film flow

presented the contact angle dependence of a dimensionless transition film height $\tilde{h}_t(\theta)$ where the influences of both effects on the flow rate just cancel each other.

The influence of side walls on the free surface shape has been investigated, one, for the steady case and two, for a slowly draining flow. We have found, that the influence of the side wall presence is observable only at distances of the order of the generalized capillary length. In the slowly draining case we have found deviations of the free surface shape from the steady case, which cannot be described by means of a dynamic contact angle which is smaller than the static contact angle. Thus, we conclude, that such a slowly draining flow cannot be modeled by a series of quasi-steady states, although the settling speed of the draining flow is many orders of magnitudes smaller than the mean flow velocity. Numerical simulations of the draining flow case revealed an indentation of the free surface near the side wall, which might promote film rupture in industrial thin film applications.

Many industrial processes deal with the problem of thin liquid films and the side wall effects involved due to the finite width of the apparatus. In coating applications for example one usually cuts away the near-wall region of the coated substrate because of its nonuniform thickness[6]. With this work we provide tools to optimize the flow rate and contact angle settings in terms of an economic utilization of the coating substance and an optimal uniformity of the coating thickness.

4.2 Stability near the side walls

4.2.1 Results

The primary convective instability of a free surface channel flow with side walls, which is characterized in section 4.1, has been studied experimentally as described in section 2.4.5. The studied liquid was Silicone oil BC50 from *Basildon* (see section 2.1).

Figure 4.14 shows the neutral stability curves for two different contact angles θ and six different distances from the side wall. Lines between the measured points are linear interpolations to guide the eye. The dashed line corresponds to the neutral stability curve for the plane flow of infinite extend. The side wall distances d_s of the measurement points range from 85 mm, which corresponds to the channel center, to 5 mm. A further reduction of the side wall distance was not possible due to the strong curvature of the free surface in the vicinity of the side wall. For side wall distances below 5 mm the laser beams became too strong distorted at the reflection point on the free surface. The excitation frequency f_e was varied between ~ 1.8 Hz and ~ 6.5 Hz. A slower excitation typically resulted in very strong peaks of higher harmonics of the fundamental excitation frequency in the Fourier transformed measurement signals (see Figures 2.15(a), 2.15(b) and 2.15(c)). The highest recordable frequency was limited by the capturing rate of the cameras.[85]

In the studied system the dimensionless film height $\tilde{H} = H/L$ was always close to one. Therefore, the influences of the retarding no–slip condition at the side walls and of the velocity overhoot on the volume flux, illustrated in Figure 4.4, almost cancel each other (see Figure 4.5(a)). The resulting difference between the Reynolds number of a two–dimensional flow defined as $\mathrm{Re}_{2D} = u_s H \nu^{-1}$ and the Reynolds number of the three dimensional flow defined as $\mathrm{Re} = 3\dot{V}/(2\nu B)$ is less than 1%. Hence we have measured the film height in the middle of the channel H with a micrometer screw as described in section 2.4.1 to determine the Reynolds number and omit the distinction between Re_{2D} and Re in the following.

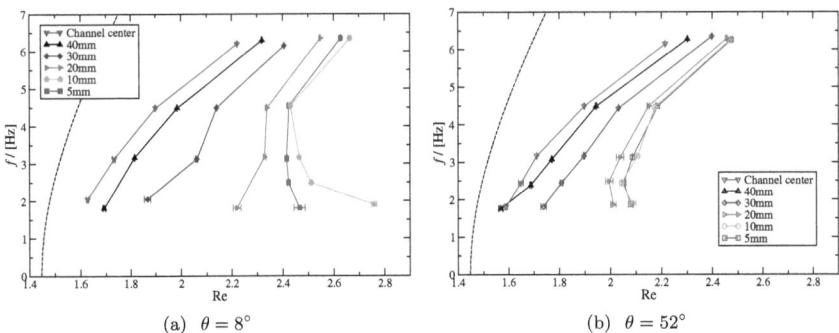

(a) $\theta = 8°$ (b) $\theta = 52°$

Figure 4.14: Neutral stability curves for different side wall distances d_s and contact angles θ. The dashed line indicates the neutral stability curve for a channel of infinite extent $B \to \infty$. Reprinted with permission from [70]. ©2011, American Institute of Physics.

In the middle of the channel the longest waves turn out to be the most unstable ones as it is the case in absence of side walls[46, 47]. However, quantitatively we find the

Chapter 4. Three–dimensional film flow

studied system to be more stable in the middle of the channel than the two–dimensional case, especially at high excitation frequencies. The neutral curve in the middle of the channel shows no contact angle dependence.

Vlachogiannis et al. [68] and Georgantaki et al. [69] studied the influence of a finite channel width on the stability of the flow at different inclination angles and fluids properties at very low excitation frequency ($f_e = 0.167\,\mathrm{Hz}$). They also found the flow to be more stable under the presence of side walls, but only if the Kapitza number Ka is sufficiently high (see Figure 4.15). The Kapitza number is defined as $\mathrm{Ka} = \sigma/(\rho g^{1/3} \nu^{4/3})$ and represents the ratio of capillary stresses to viscous stresses. It is a dimensionless material property only and does not depend on flow properties. They found the ratio $R^* = \mathrm{Re}_c/\mathrm{Re}_{c,2D}$ to depend only on the channel width and on the Kapitza number, but not on the channel inclination or material properties like for example the surface tension σ.

When the Kapitza number is of the order of one or smaller they found no influence of the channel width on the stability of the flow when the channel was at least 100 mm broad. Although all our measurements were done at a Kapitza number of approximately five we found a strong stabilizing influence of the side walls on the flow which is not in line with the data shown in Figure 4.15. Yet, we suppose that these findings are not in conflict with each other, because an extrapolation of our data shown in Figure 4.14 yields that it might coincide with the two–dimensional case at the limit of very low frequency $f_e \to 0$ as proposed by the work of Georgantaki et al. [69].

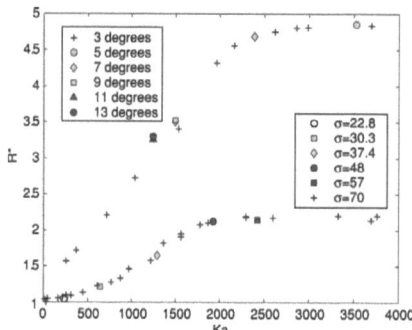
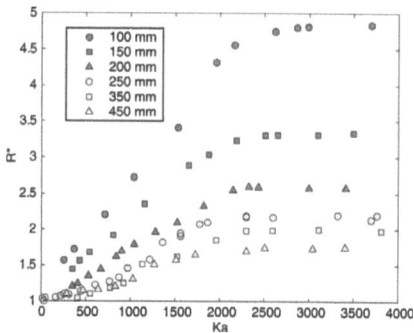

(a) Ratio R^* as a function of Ka for several inclination angles, fluids and channel widths. The dimension of the given surface tension is mN/m. The upper curve corresponds to a channel width of $B = 100\,\mathrm{mm}$. The lower curve corresponds to a channel width of $B = 250\,\mathrm{mm}$.

(b) Ratio R^* as a function of Ka for several channel widths B, $\alpha = 3\,°$.

Figure 4.15: Ratio R^* as a function of Ka. Reprinted with permission from [69]. ©2011, American Physical Society. URL: http://pre.aps.org/abstract/PRE/v84/i2/e026325

In the vicinity of the side walls all our measurements showed a further stabilization of the flow compared to the data from the center–line measurements. Furthermore, the contact angle θ, which did not play a role in the middle of the channel, gains in importance when the side wall distance d_s is reduced. Measurements done in the channel with the coated glass side walls ($\theta = 52\,°$) show neutral curves which are significantly shifted

4.2. Stability near the side walls

to lower Reynolds numbers than the measurements done in the channel with untreated Plexiglas® side walls ($\theta = 8°$). We account two effects for this phenomenon. One, the smaller contact angle causes a larger capillary elevation height Δh (see e.g. Figure 4.10 or equation (4.27)) and thus to a larger contact area between the liquid and the retarding side wall. Two, when a finite surface tension of the liquid is considered, the smaller contact angle leads to a stronger curvature and therefore to a stronger pretensioning of the free surface in the vicinity of the side wall which hinders free surface waves to develop and thus tends to stabilize the flow.

We find a remarkable range of the contact angle influence on the stability of the flow. At a side wall distance of 10 mm, which is about four times the capillary length L or the film height H, the difference between the neutral curves at $\theta = 8°$ and $\theta = 52°$ is up to 25%. Even up to a side wall distance of 40 mm, which is about 17 times the capillary length L or the film height H, the difference between the neutral curves at $\theta = 8°$ and $\theta = 52°$ is still more than 7%.

Additionally, the shape of the neutral curves changes when the side wall distance is decreased. In the middle of the channel we observe the longest waves to become initially unstable as predicted by Benjamin[46] and Yih[47] for the two–dimensional case. For side wall distances of 10 mm in the case of uncoated side walls and 20 mm in the case of coated side walls the type of the instability changes from a long–wave type to a short–wave type instability in the investigated frequency range. This type of instability is well known for boundary layer flows as observed experimentally by Schubauer and Skramstad[86] for a plate which is aligned parallel to a plane flow. This configuration was later described in detail by Schlichting and Gertsen[87]. The mechanisms for the instability are quite different because in the present work an instability of a free surface near a side wall and not the instability of a bulk is investigated. Typical critical Reynolds numbers found for this bulk instability in a boundary layer are about two orders of magnitude larger than for the free surface flow investigated here[87]. However the similarity of the shape of the neutral curves close to the side wall suggests to treat the near wall region as a capillary boundary layer with a range of four to eight times the capillary length L.

Other gravity–driven free surface flows showing a short–wave instability are described in a two–dimensional theoretical framework by D'Alessio et al.[59] for Newtonian liquids at very high inverse Bond numbers, which means that capillary forces dominate over gravity, or by Heining and Aksel[60] for power–law liquids flowing down a sinusoidally undulated incline.

At intermediate side wall distances the neutral curves neither show the character of a typical long–wave instability nor the typical short–wave instability. In this transition region we observe, that the neutral curves have an inflection point in the investigated frequency range. The size of the transition region seems to be larger for smaller contact angles θ.

Figure 4.16 shows the side wall distance dependence of the neutral points for both investigated contact angles and two different excitation frequencies. Because the measurements were done at slightly different excitation frequencies the data shown in Figure 4.14 have been interpolated linearly to provide comparability. In the middle of the channel we do not observe a contact angle dependence as a comparison of the Figures 4.14(a) and 4.14(b) already revealed. Reducing d_s leads at first to a monotonous increase of the Reynolds number at which free surface waves are neither damped, nor amplified while travelling downstream, which means, that the flow is getting more and more stable due to the retarding influence of the side wall and the pretensioning of the free surface coming into play. Especially at low excitation frequency f_e we observe the large amplitude and range of the contact angle influence.

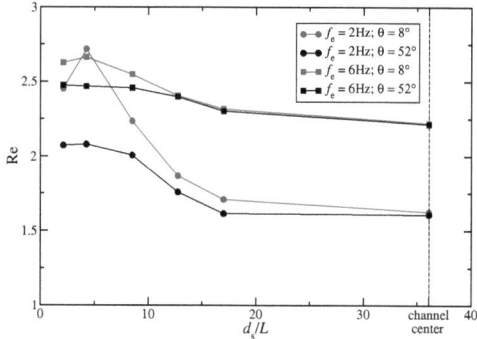

Figure 4.16: Side wall distance dependence of the neutral points at two different contact angles θ and excitation frequencies f_e. Reprinted with permission from [70]. ©2011, American Institute of Physics.

However, this monotonous behavior is broken very close to the side wall, although the stabilizing effects of the side wall should be strongest here. This can also be seen in Figure 4.14: Especially in the case of uncoated side walls the neutral curve for $d_s = 5\,\text{mm}$ is left of the neutral curve for $d_s = 10\,\text{mm}$ over the whole investigated excitation frequency range. Obviously another (competing) effect, namely the presence of a velocity overshoot, gains in importance there. When the film thickness is smaller than the critical film thickness h_c, which is about $\sim 1.3L$ for small contact angles θ, not only the film thickness in the vicinity of the side wall is larger than the film thickness in the middle of the channel H, but also a velocity overshoot is observed due to the capillary elevation (see section 4.1). Both, the higher film thickness as well as the higher velocity at the free surface cause the local Reynolds number $\text{Re}_{\text{loc}}(d_s) = h(d_s)u_s(d_s)/\nu$ to exceed the (global) Reynolds number $\text{Re} = 3\dot{V}/(2\nu B)$ at some d_s (See Figure 4.17). In those regions the onset of waves at the free surface is promoted and the flow tends to be more unstable. Due to the more pronounced velocity overshoot and capillary elevation for smaller contact angles the magnitude of the local Reynolds number overshoot is larger for $\theta = 8\,°$. That explains why the peaks in the side wall distance dependence of the neutral points in Figure 4.16 are more pronounced for $\theta = 8\,°$ than for $\theta = 52\,°$. Compared to the stabilizing effects, the destabilizing influence of the local Reynolds number overshoot seems to be of a much shorter range.

4.2.2 Conclusions

We have shown that the neutral curve for the onset on a primary instability in gravity-driven free surface flows depends strongly on the distance to the side wall of the channel. In the studied system the flow in the vicinity of the side wall was always more stable

4.2. Stability near the side walls

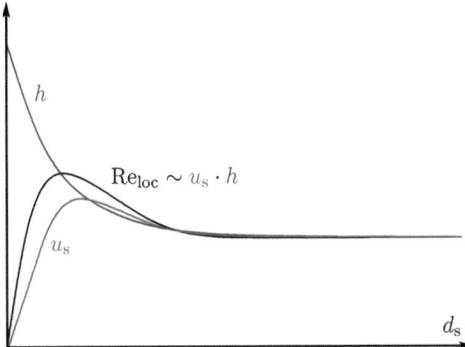

Figure 4.17: Sketch of the side wall distance dependence of the local Reynolds number Re$_{loc}$ when a velocity overshoot is present.

than the flow in the middle of the channel. A direct consequence is, that a flow may be unstable at some regions, but shows other regions where the free surface waves coming from the unstable regions are damped at the same time. Of course a flow has to be treated as unstable as soon as first surface waves appear somewhere for most applications. Nevertheless, we found that a coexistence of stable and unstable regions is possible, which is important to understand the influence of the side walls on the instability of the flow and the underlying mechanisms responsible.[70]

In the middle of the channel we observe a long–wave type instability as found by Benjamin[46] and Yih[47] for a purely two–dimensional flow. When the distance to the side wall is reduced the instability undergoes a transition from the long–wave type to a short–wave type as it is typical for boundary layer flows[86, 87].

Georgantaki et al.[69] restricted their studies on the influence of the channel width on the instability of film flow to the limit of very long waves. They find the ratio R^*, which is the critical Reynolds number normalized with the critical Reynolds number of a film flowing down a plane of infinite extent, to be a function of the channel width and the Kapitza number. Since we have shown, that the most unstable wave may also have a finite wavelength in near wall regions, we propose that a excitation frequency variation has to be carried out to determine the ratio R^* properly for all channel widths and Kapitza numbers.

One has to consider different competing effects of the side walls on the instability of the free surface, some tend to stabilize and some tend to destabilize the flow. The additional no slip condition at the wall and the pretensioning of the free surface due to capillary elevation tend to stabilize the flow. These effects are more pronounced when the contact angle between the liquid and the side wall θ is small. For thin films the capillary elevation leads to a velocity overshoot at the free surface and thus to an overshoot of the local Reynolds number which promotes the onset of free surface waves.

Compared to the effects which tend to stabilize the flow, which are still significant up to side wall distances of $17L$, the influence of the local Reynolds number overshoot on the stability of the free surface seems to be of rather short range.

Also the magnitude of the impact of the Reynolds number overshoot in the vicinity of the side wall is clearly of minor importance compared to the stabilizing effects for the investigated flow configuration. But we remark that a further reduction of the film height will amplify the magnitude of the Reynolds number overshoot until its influence might overcome the stabilizing effects. Therefore, we speculate that there should exist a critical film height h'_c at which a flow initially becomes unstable close to the side walls before the Reynolds number in the middle of the channel reaches the classical result of $\text{Re}_c = (5/4)\cot\alpha$. This would imply that a film of thickness $h < h'_c$ which is confined by side walls initially becomes unstable at a smaller Reynolds number than a channel flow of infinite extent ($B \to \infty$) with the same film thickness h. To prove this assumption that the presence of side walls have an overall destabilizing influence on very thin gravity–driven film flows further experiments have to be carried out.

Chapter 5
Conclusions and outlook

In the present work we study viscous gravity–driven film flow down an inclined channel numerically, analytically and experimentally. Particular focus lies on the influence of a periodic two–dimensional sinusoidally undulated topography on the flow and the appearance and disappearance of eddies in the valleys of the undulation. The corresponding results are presented in chapter 3. Chapter 4 deals with the question how the presence of side walls and different contact angles between the liquid and the side wall influences the flow structure, the overall volume flux and the physical stability of the flow. The main findings are summarized in the following.

First, we consider numerically and experimentally a steady flow down an inclined periodic two–dimensional sinusoidally undulated topography. We find, that the free surface undergoes a sharp transition from a strongly anharmonic shape with a strong indentation just above the valleys of the undulation to a smooth harmonic shape as the Reynolds number is increased above a transition Reynolds number. At this transition Reynolds number a hydraulic jump which has been established continuously with increasing Reynolds number disappears abruptly because the flow changes its type from a sub- to a supercritical flow in terms of the Froude number. Furthermore, we find that eddies, which are formed with increasing Reynolds number in the valleys of the underlying undulation, disappear again just in the vicinity of the transition Reynolds number. However, the eddies are suppressed when the hydraulic jump is present below the transition Reynolds number as well as above it. Hence, we conclude, that not the transition itself is responsible for the suppression of the eddy structures, but an amplification of the free surface amplitude, which comes along with the transition and is well known as a resonance phenomenon in literature.

Such systematic suppression of eddies is of particular relevance to open up new optimized process windows for many industrial applications where their formation is desired or undesired, depending on the purpose, since they have significant influence on macroscopic system properties. In heat exchanger applications for example the formation of such eddy structures has a major impact on the convective heat transport in the liquid while they lead to drag reduction, which might be useful for bearings or any kind of material transport. In environmental systems particles captured in the recirculating flow are cut off from subsequent delivery of potentially necessary nutrient substances and so is the wall which is in contact with the eddy. The possibility to generate or destroy eddy formations in the valleys of an undulation just by damping or exciting resonance seems

Chapter 5. Conclusions and outlook

to be a good framework in general, different from applying external forces to the system.

In the second part of the present work the influence of side walls and the contact angle between the liquid there on the velocity profile and the flow rates has been studied experimentally and theoretically. In the case of a wetting liquid a capillary elevation is formed at the side walls, which may lead to a velocity overshoot in the vicinity of the wall. While this velocity overshoot leads to an increase of the transported liquid in the channel an additional no–slip condition at the side walls has a contrary influence on the flow rate. Here, a criterion for the first onset of a velocity overshoot is presented, as well as a criterion where the influences of the velocity overshoot and of the additional no–slip condition on the overall volume flux just cancel each other. As long as they do not cancel each other, neglecting the three–dimensionality of the flow would lead to an over– or underestimation of the flow rate, what becomes especially important in the case of a very thin film, where most liquid is transported in the capillary corner of the channel.

In addition to the steady flow case findings for the case of a slowly draining flow are presented. Experiments show, that the free surface shape of the liquid can not be modeled by a series of quasi–steady states, even if a dynamic contact angle is taken into consideration, although the settling speed of the liquid is orders of magnitude smaller than the mean flow velocity. Time dependent numerical simulations of the slowly draining flow case revealed an indentation of the free surface near the side walls, which might promote film rupture in industrial thin film applications.

Finally, we investigate experimentally the influence of the presence of side walls on the primary instability of the free surface of the flow and find, that the neutral stability curve shows a strong dependence on the distance to the side wall and on the wetting properties of the liquid. In the middle of the channel, far away from the side walls, we observe a long–wave type instability of the free surface, which is independent of the contact angle between the liquid and the side wall. However, when the side wall distance is reduced the type of the instability changes to a short–wave type instability, as it is well known from boundary layer flows at much higher Reynolds numbers, and shows a strong contact angle dependence. We find, that one has to consider different competing effects of the side walls on the stability of the flow. The pretensioning of the free surface due to the capillary elevation and the additional no–slip boundary condition at the side wall tend to stabilize the flow and are sensible up to large side wall distances of about 17 times the generalized capillary length. The formation of a velocity overshoot in the capillary elevation, on the contrary, tends to destabilize the flow, because the local Reynolds number in the near wall region can exceed the (global) Reynolds number of the channel flow. However, compared to the stabilizing effects coming from the side wall the destabilizing effect of the side walls seems to be of rather short range and plays a minor part only for the investigated flow configuration. Thus, the presence of side walls leads to an overall stabilization of the system studied here. Nevertheless, we speculate that for even thinner films the increasing magnitude of the velocity overshoot might cause the destabilizing effect to win over the stabilizing effects, which would implicate that very thin films bounded by side walls would be more unstable than the corresponding films of infinite extent.

In industrial applications which deal with very thin films like curtain coating processes for example, one would like to optimize the process in terms of a fast processing and an effective exploitation of the coating material. Since side walls lead to a nonuniform coating at the edges one usually cuts the edges of the coated material away to get a uniformly coated result. Here, we show that especially in thin films most of the material is transported just where edges are located when the contact angle of the liquid at the side wall is below 90°. The speed of processing in industrial applications is often limited by the stability of the flow, which is also strongly affected by the side walls and the wetting properties of the liquid. Therefore, we suggest that the contact angle should be considered for the design of such devices since it holds some relevant potential for optimization in coating and other thin film applications.

To summarize, the present work deals with the influence of a corrugated topography and of side walls on film flows. Nevertheless, several aspects concerning the stability of gravity–driven film flows remain open questions. Since our experiments on the influence of the side wall on the stability of the flow were restricted to side wall distances of 5 mm, we suggest a full numerical study of the Navier–Stokes equations involving extensive parameter studies especially for the case of very thin films to get a closer look into the liquid near the side wall and thus a better understanding of the underlying physics. However, we remark, that the computational effort for such simulations will quickly become a key issue, since for the investigation of convective instability it is necessary to simulate rather large domains which are fully three–dimensional and time dependent on a rather precise mesh to resolve capillary effects properly. In addition to the question how side walls influence the flows' stability, the flow over substrates with finite corrugations has been studied by several authors over the last years theoretically as well as experimentally. However, the question how a formation of eddies in steep substrate undulations might influence the stability of the flow is very interesting but still open and is worth to be addressed in future works.

Chapter 5. Conclusions and outlook

List of symbols

(x, y, z)	spatial coordinates
$(\tilde{x}, \tilde{y}, \tilde{z})$	dimensionless spatial coordinates
$(\hat{e}_x, \hat{e}_y, \hat{e}_z)$	unit vectors in x-, y- and z-direction
(X, Y, Z)	spatial coordinates in the reference frame of the traverse
(x', y', z')	spatial coordinates in the reference frame of the screen
(u, v, w)	velocity components in x-, y- and z-direction
$(\tilde{u}, \tilde{v}, \tilde{w})$	dimensionless velocity components in x-, y- and z-direction
$\mathbf{u}$	velocity vector
t	time
ρ, ν, η, σ	fluid properties: density, kinematic viscosity, dynamic viscosity and surface tension
$\mathbf{n}$	outer normal unit vector
$\underline{\mathbf{T}}$	stress tensor
$\underline{\mathbf{I}}$	identity matrix
$g, \tilde{g}$	dimensional and dimensionless acceleration of gravity
$\mathbf{g}, \tilde{\mathbf{g}}$	dimensional and dimensionless gravity vector
$p, \tilde{p}$	dimensional and dimensionless pressure
$p_0, \tilde{p}_0$	dimensional and dimensionless ambient pressure
a, λ	amplitude and wavelength of the channel undulation
α	channel inclination angle
B	channel width
$b, \tilde{b}$	dimensional and dimensionless contour of the channel topography
l_1, l_2	laser beam travel distances
$h, \tilde{h}$	dimensional and dimensionless position of the free surface
$d, \tilde{d}$	dimensional and dimensionless film thickness
$h_\mathrm{n}, d_\mathrm{n}$	Nusselt film thickness
$H, \tilde{H}$	dimensional and dimensionless film height in the middle of the channel
H_m	measurement height
u_s	free surface velocity
$u_\mathrm{s,m}$	free surface velocity in the middle of the channel
u_max	maximal free surface velocity
$\bar{u}$	mean flow velocity
$\bar{u}_\mathrm{n}$	mean velocity of a Nusselt film flow
u_loc	local free surface velocity
y_max	y-location of the maximal free surface velocity

Chapter 5. Conclusions and outlook

Re	(global) Reynolds number
$Re_{2D}, Re_{(a)}$	Reynolds number for an infinite broad channel
$Re_{(b)}$	Reynolds number of a flow with side walls without capillary elevation
$Re_{(d)}$	Reynolds number of the capillary corner flow
Re_{loc}	local Reynolds number
Re_s	Reynolds number for the settling of a draining film
Re_c	critical Reynolds number
$Re_{c,2D}$	critical Reynolds number for a plane channel of infinite extent
Re_1, Re_2, Re_3	Reynolds numbers where an eddy appears or disappears
R^*	normalized critical Reynolds number
Ka	Kapitza number
Bo	Bond number
Fr	Froude number
$\dot{q}, \dot{V}$	flow rate, volume flux
$\dot{V}_{2D}$	volume flux of a two dimensional flow without side–walls
$\dot{V}_{\theta=90°}$	volume flux of a flow with side–walls but without capillary elevation
θ	static contact angle between liquid and side–walls
θ_d	dynamic contact angle between liquid and side–walls
T	temperature
u_{sed}	sedimentation speed
r_s, ρ_s	sphere radius and density
f	frequency
f_e	excitation frequency
f_{min}	minimal excitation frequency
$\dot{V}_{max}$	maximal volume flux
Re_{max}	maximal Reynolds number
h_{max}	maximal film thickness
λ_{max}	maximal wavelength
$u_{s,max}$	maximal free surface velocity
d_s	side–wall distance
$\mathbf{p}_i(t), \ i \in \{1,2\}$	laser spot positions of laser 1 and laser 2
$p_i(t), \ i \in \{1,2\}$	absolute values of the laser spot positions of laser 1 and laser 2
$p_{x',i}, p_{y',i}, \ i \in \{1,2\}$	x'- and y'-positions of laser 1 and laser 2
$\hat{\mathbf{p}}_i(f), \ i \in \{1,2\}$	Fourier transformed laser spot positions of laser 1 and laser 2
$\hat{p}_i(f), \ i \in \{1,2\}$	absolute values of the Fourier transformed laser spot positions of laser 1 and laser 2
$\hat{p}_{x',i}, \hat{p}_{y',i}, \ i \in \{1,2\}$	Fourier transformed x'- and y'-positions of laser 1 and laser 2
$\delta p_i, \ i \in \{1,2\}$	deflection of laser 1 and laser 2 on the screen
$\Delta \hat{p}_{12}$	difference of peak heights of the signals $\hat{p}_1$ and $\hat{p}_2$
$\Delta h, \Delta \tilde{h}$	dimensional and dimensionless height of the capillary elevation at the side–walls
$\zeta(y)$	capillary film elevation
L, l	capillary length, capillary range

$\tilde{h}_t, \tilde{h}_c$	dimensionless transition film thickness, dimensionless critical film thickness
h'_c	critical film height
r	combined dimensionless parameter
r_c	critical ratio
κ	curvature of the free surface
R	radius of curvature of the free surface
ξ	dimensionless steepness parameter
k	dimensionless wave number
p_r	reference pressure
c_1, c_2	fit parameters
$m_{\max}$	maximal slope of a travelling free surface wave
A	amplitude of a travelling free surface wave
λ_w	wavelength of a travelling free surface wave
D_n	integration constants
c_n	Fourier coefficient constants for particular solution of the velocity field
d_m	inhomogeneity of the set of algebraic equations
E_n	integration constants
G_{nm}	matrix for set of algebraic equations
ξ_d	grade of dispersity of the particle size distribution
x_{50}	median particle size
x_{16}, x_{84}	particle sizes

List of Figures

2.1 Sketch of the flow circuit including the channel which is mounted on a vibration isolating table and a pump which transports the liquid from a large liquid reservoir to a smaller inflow tank on top of the channel. . . . 17
2.2 Geometry of the two–dimensional undulated inlay. 18
2.3 Particle size distributions. 19
2.4 Emission spectrum of *Red Fluorescent Polymer Microspheres* from *Duke Scientifics* in silicone oil. 19
2.5 Illustration of the tip of a needle which is less than 6.5 μm above the surface of a flowing liquid film and just touching it. 20
2.6 Experimental setup for the surface contour detection and the streamline detection in the troughs of the undulated inlay in channel 1. 22
2.7 Photo of the experimental setup for the surface contour detection and the streamline detection in the troughs of the undulated inlay in channel 1. . 22
2.8 Measurement setup for the free surface shape detection in channel 2. . . . 23
2.9 Illustration of the evaluation method for the reconstruction of the shape of the underlying topography and the streamline pattern from experimental single image data. 24
2.10 Sketch of the experimental setup for the free surface stability measurements. 25
2.11 Photo of the experimental setup for the measurements of the free surface stability. 26
2.12 Channel top view illustrating the measurement positions. 27
2.13 Recorded positions of both laser spots during an experimental run. 27
2.14 Section of the time dependence of the x'- and y'-components of both laser spots during an experimental run. 28
2.15 Absolute values of the Fourier transformed signals of both lasers at different Reynolds numbers and their amplitude differences. 29

3.1 Viscous film flow down a wavy incline. 32
3.2 Comparison of experimental path lines to numerical streamlines. 35
3.3 Illustration of the averaging process for the detection of the free surface shape. 36
3.4 Comparison of experimental and numerical free surface shapes. 37
3.5 Cross–sectional area of the eddy as a function of the Reynolds number at different inclination angles. 37
3.6 Mean film thickness averaged over one bottom period. 38
3.7 Amplitude of the first harmonic of the free surface shape. 39

List of Figures

3.8 Amplitude of the second harmonic of the free surface shape. 39
3.9 Average film thickness and amplitude of the first two Fourier components. 40
3.10 Snapshot of a video which illustrates the free surface shape transition at an intermediate Reynolds number. 41
3.11 Spatial dependence of the local Froude number. 42
3.12 Comparison of the position of the free surface shape transition and the eddy–free window position at different channel inclinations. 43
3.13 Numerically observed streamline patterns by Trifonov[37]. 45

4.1 Channel geometry illustrating side wall effects on the flow. 47
4.2 Cross sectional velocity profiles of different channel flow types. 51
4.3 Decrease of the volume flux due to the additional no–slip condition at the side walls (without capillarity) compared to the two–dimensional case. . . 52
4.4 Free surface velocity profile in flow direction showing velocity overshoot and defect compared to the plane flow with the same film height. 53
4.5 Influence of capillary effects at the side walls on the normalized flow rate. 53
4.6 Dependence of the transition film thickness $\tilde{h}_\mathrm{t}$ on the contact angle θ. . . 54
4.7 Comparison of measured and calculated velocity profiles for different film heights H, measurement heights H_m and contact angles θ. 55
4.8 Influence of the film thickness on the velocity overshoot. 56
4.9 Dependence of the critical film thickness $\tilde{h}_\mathrm{c}$ on the contact angle θ. 57
4.10 Theoretical and experimental shape of the free surface in the vicinity of the side wall. 57
4.11 Comparison of the steady free surface shape with the free surface shape of the draining flow 13.3 s after switching off the pump. 58
4.12 Evolution of the free surface position in the middle of the channel and of the near wall region after switching off the pump. 59
4.13 Numerical simulations of the free surface shape of a draining flow over one half of the channel $0 \leq d_\mathrm{s} \leq B/2$. 60
4.14 Neutral stability curves for different side wall distances d_s and contact angles θ. 62
4.15 Ratio R^* as a function of Ka. 63
4.16 Side wall distance dependence of the neutral points at two different contact angles θ and excitation frequencies f_e. 65
4.17 Sketch of the side wall distance dependence of the local Reynolds number Re_loc when a velocity overshoot is present. 66

Bibliography

[1] Greve, R. and Blatter, H. Dynamics of Ice Sheets and Glaciers. Springer, Berlin, 1st edition, (2009).

[2] Luca, I., Hutter, K., Tai, Y., and Kuo, C. A hierarchy of avalanche models on arbitrary topography. Acta Mechanica **205**, 121–149 (2009).

[3] Hutter, K., Svendsen, B., and Rickenmann, D. Debris flow modeling: A review. Continuum Mechanics and Thermodynamics **8**(1), 1–35 (1994).

[4] Kistler, S. F. and Schweizer, P. M. Liquid Film Coating. Chapman and Hall, New York, (1997).

[5] Weinstein, S. J. and Ruschak, K. J. Coating flows. Annual Review of Fluid Mechanics **36**(1), 29–53 (2004).

[6] Gilbert, G., Robert, B., and Mauron, M. Operative limits of curtain coating due to edges. Chemical Engineering and Processing: Process Intensification **50**, 462–465 (2011).

[7] Webb, R. L. Principles of Enhanced Heat Transfer. Wiley, New York, (1994).

[8] Vlasogiannis, P., Karagiannis, G., Argyropoulos, P., and Bontozoglou, V. Air-water two-phase flow and heat transfer in a plate heat exchanger. International Journal of Multiphase Flow **28**(5), 757–772 (2002).

[9] Nusselt, W. Die Oberfächenkondensation des Wasserdampfes. VDI Zeitschrift **60**, 541–546 (1916).

[10] Kapitza, P. L. and Kapitza, S. P. Wave flow of thin layers of a viscous liquid. Zh. Eksp. Teor. Fiz **18**, 3–28 (1948).

[11] Kapitza, P. L. Wave flow of thin viscous liquid layers. Zh. Eksp. Teor. Fiz **19**, 3–28 (1948).

[12] Chang, H. C. Wave evolution on a falling film. Annu. Rev. Fluid Mech **26**, 103–136 (1994).

[13] Demekhin, E., Shapar, E., and Selin, A. Surface instability of the liquid turbulent flows in the open inclined channels. Thermophysics and Aeromechanics **14**, 223–230 (2007). 10.1134/S0869864307020084.

[14] Decré, M. M. J. and Baret, J.-C. Gravity-driven flows of viscous liquids over two-dimensional topographies. Journal of Fluid Mechanics **487**, 147–166 (2003).

[15] Hayes, M., O'Brien, S. B. G., and Lammers, J. H. Green's function for steady flow over a small two-dimensional topography. Physics of Fluids **12**(11), 2845–2858 (2000).

[16] Kalliadasis, S., Bielarz, C., and Homsy, G. M. Steady free-surface thin film flows over topography. Physics of Fluids **12**(8), 1889–1898 (2000).

[17] Mazouchi, A. and Homsy, G. M. Free surface Stokes flow over topography. Physics of Fluids **13**(10), 2751–2761 (2001).

[18] Aksel, N. Influence of the capillarity on a creeping film flow down an inclined plane with an edge. Archive of applied Mechanics **70**, 81–90 (2000).

[19] Negny, S., Meyer, M., and Prevost, M. Study of a laminar falling film flowing over a wavy wall column: Part i. numerical investigation of the flow pattern and the coupled heat and mass transfer. International Journal of Heat and Mass Transfer **44**(11), 2137–2146 (2001).

[20] Wierschem, A., Scholle, M., and Aksel, N. Comparison of different theoretical approaches to experiments on film flow down an inclined wavy channel. Experiments in Fluids **33**(3), 429–442 (2002).

[21] Wierschem, A., Lepski, C., and Aksel, N. Effect of long undulated bottoms on thin gravity-driven films. Acta Mechanica **179**(1), 41–66 (2005).

[22] Trifonov, Y. Y. Viscous film flow down corrugated surfaces. Journal of Applied Mechanics and Technical Physics **45**, 389–400 (2004). 10.1023/B:JAMT.0000025021.41499.e1.

[23] Luo, H. and Pozrikidis, C. Gravity-driven film flow down an inclined wall with three-dimensional corrugations. Acta Mechanica **188**(3), 209–225 (2007).

[24] Sadiq, I. M. R., Gambaryan-Roisman, T., and Stephan, P. Falling liquid films on longitudinal grooved geometries: Integral boundary layer approach. Physics of Fluids **24**(1), 014104 (2012).

[25] Heining, C., Pollak, T., and Aksel, N. Pattern formation and mixing in thee-dimensional film flow. Physics of Fluids **in press** (2012).

[26] Sellier, M. Substrate design or reconstruction from free surface data for thin film flows. Physics of Fluids **20**(6), 062106 (2008).

[27] Heining, C. and Aksel, N. Bottom reconstruction in thin-film flow over topography: Steady solution and linear stability. Physics of Fluids **21**(8), 083605 (2009).

[28] Sellier, M. and Panda, S. Beating capillarity in thin film flows. International Journal for Numerical Methods in Fluids **63**(4), 431–448 (2010).

Bibliography

[29] Heining, C. Velocity field reconstruction in gravity-driven flow over unknown topography. Physics of Fluids **23**(3), 032101 (2011).

[30] Heining, C., Sellier, M., and Aksel, N. The inverse problem in creeping film flows. Acta Mechanica **223**, 1–7 (2012).

[31] Pozrikidis, C. The flow of a liquid film along a periodic wall. Journal of Fluid Mechanics Digital Archive **188**, 275–300 (1988).

[32] Scholle, M., Rund, A., and Aksel, N. Drag reduction and improvement of material transport in creeping films. Arch. Appl. Mech **75**, 93–112 (2006).

[33] Wierschem, A., Scholle, M., and Aksel, N. Vortices in film flow over strongly undulated bottom profiles at low Reynolds numbers. Physics of Fluids **15**(2), 426–435 (2003).

[34] Scholle, M., Wierschem, A., and Aksel, N. Creeping films with vortices over strongly undulated bottoms. Acta Mechanica **168**(3), 167–193 (2004).

[35] Wierschem, A. and Aksel, N. Influence of inertia on eddies created in films creeping over strongly undulated substrates. Physics of Fluids **16**(12), 4566–4574 (2004).

[36] Scholle, M., Haas, A., Aksel, N., Wilson, M. C. T., Thompson, H. M., and Gaskell, P. H. Competing geometric and inertial effects on local flow structure in thick gravity-driven fluid films. Physics of Fluids **20**(12), 123101 (2008).

[37] Trifonov, Y. Y. Viscous liquid film flows over a periodic surface. International Journal of Multiphase Flow **24**(7), 1139–1161 (1999).

[38] Zhao, L. and Cerro, R. Experimental characterization of viscous film flows over complex surfaces. International Journal of Multiphase Flow **18**(4), 495–516 (1992).

[39] Bontozoglou, V. and Papapolymerou, G. Laminar film flow down a wavy incline. International Journal of Multiphase Flow **23**(1), 69–79 (1997).

[40] Bontozoglou, V. Laminar film flow along a periodic wall. Computer Modelling in Engineering & Sciences **1**, 133–142 (2000).

[41] Wierschem, A. and Aksel, N. Hydraulic jumps and standing waves in gravity-driven flows of viscous liquids in wavy open channels. Physics of Fluids **16**(11), 3868–3877 (2004).

[42] Wierschem, A., Bontozoglou, V., Heining, C., Uecker, H., and Aksel, N. Linear resonance in viscous films on inclined wavy planes. International Journal of Multiphase Flow **34**(6), 580–589 (2008).

[43] Heining, C., Bontozoglou, V., Aksel, N., and Wierschem, A. Nonlinear resonance in viscous films on inclined wavy planes. International Journal of Multiphase Flow **35**(1xperim), 78–90 (2009).

[44] Wierschem, A., Pollak, T., Heining, C., and Aksel, N. Suppression of eddies in films over topography. Physics of Fluids **22**, 113603 (2010).

[45] Nguyen, P. K. and Bontozoglou, V. Steady solutions of inertial film flow along strongly undulated substrates. Physics of Fluids **23**, 052103 (2011).

[46] Benjamin, T. B. Wave formation in laminar flow down an inclined plane. Journal of Fluid Mechanics **2**(06), 554–573 (1957).

[47] Yih, C.-S. Stability of liquid flow down an inclined plane. Physics of Fluids **6**(3), 321–334 (1963).

[48] Liu, J., Paul, J. D., and Gollub, J. P. Measurements of the primary instabilities of film flows. Journal of Fluid Mechanics **250**, 69–101 (1993).

[49] Liu, J. and Gollub, J. P. Solitary wave dynamics of film flows. Physics of Fluids **6**(5), 1702–1712 (1994).

[50] Vlachogiannis, M. and Bontozoglou, V. Observations of solitary wave dynamics of film flows. Journal of Fluid Mechanics **435**, 191–215 (2001).

[51] Chang, H. C. and Demekhin, E. A. Complex wave dynamics on thin films. Elsevier, Amsterdam, (2002).

[52] Craster, R. V. and Matar, O. K. Dynamics and stability of thin liquid films. Rev. Mod. Phys. **81**(3), 1131–1198 (2009).

[53] Wierschem, A. and Aksel, N. Instability of a liquid film flowing down an inclined wavy plane. Physica D: Nonlinear Phenomena **186**(3-4), 221–237 (2003).

[54] Dávalos-Orozco, L. A. Nonlinear instability of a thin film flowing down a smoothly deformed surface. Physics of Fluids **19**(7), 074103 (2007).

[55] Dávalos-Orozco, L. A. Instabilities of thin films flowing down flat and smoothly deformed walls. Microgravity Science and Technology **20**(3), 225–229 (2008).

[56] Benney, D. J. Long waves on liquid films. Journal of Mathematical Physics **45**(2), 150–155 (1966).

[57] Trifonov, Y. Y. Stability and nonlinear wavy regimes in downward film flows on a corrugated surface. Journal of Applied Mechanics and Technical Physics **48**(1), 91–100 (2007).

[58] Trifonov, Y. Y. Stability of a viscous liquid film flowing down a periodic surface. International Journal of Multiphase Flow **33**, 1186–1204 (2007).

[59] D'Alessio, S. J. D., Pascal, J. P., and Jasmine, H. A. Instability in gravity-driven flow over uneven surfaces. Physics of Fluids **21**(6), 062105 (2009).

[60] Heining, C. and Aksel, N. Effects of inertia and surface tension on a power-law fluid flowing down a wavy incline. International Journal of Multiphase Flow **36**(11-12), 847–857 (2010).

[61] Vlachogiannis, M. and Bontozoglou, V. Experiments on laminar film flow along a periodic wall. Journal of Fluid Mechanics **457**, 133–156 (2002).

Bibliography

[62] Argyriadi, K., Vlachogiannis, M., and Bontozoglou, V. Experimental study of inclined film flow along periodic corrugations: The effect of wall steepness. Physics of Fluids **18**(1), 012102 (2006).

[63] Scholle, M. and Aksel, N. An exact solution of visco-capillary flow in an inclined channel. Zeitschrift für Angewandte Mathematik und Physik **52**(5), 749–769 (2001).

[64] Hopf, L. Turbulenz bei einem Flusse. Annalen der Physik **337**(9), 777–808 (1910).

[65] Scholle, M. and Aksel, N. Thin film limit and film rupture of the visco–capillary gravity–driven channel flow. Zeitschrift für Angewandte Mathematik und Physik **54**(3), 517–531 (2003).

[66] Haas, A., Pollak, T., and Aksel, N. Side wall effects in thin gravity–driven film flow - steady and draining flow. Physics of Fluids **23**, 062107 (2011).

[67] Leontidis, V., Vatteville, J., Vlachogiannis, M., Andritsos, N., and Bontozoglou, V. Nominally two-dimensional waves in inclined film flow in channels of finite width. Physics of Fluids **22**(11), 112106 (2010).

[68] Vlachogiannis, M., Samandas, A., Leontidis, V., and Bontozoglou, V. Effect of channel width on the primary instability of inclined film flow. Physics of Fluids **22**(1), 012106 (2010).

[69] Georgantaki, A., Vatteville, J., Vlachogiannis, M., and Bontozoglou, V. Measurements of liquid film flow as a function of fluid properties and channel width: Evidence for surface-tension-induced long-range transverse coherence. Phys. Rev. E **84**(2), 026325 (2011).

[70] Pollak, T., Haas, A., and Aksel, N. Side wall effects on the instability of thin gravity-driven films – From long-wave to short-wave instability. Physics of Fluids **23**, 094110 (2011).

[71] Tropea, C., Yarin, A., and Foss, J. F., editors. Springer Handbook of Experimental Fluid Mechanics, p. 681. Springer, Berlin, (2007).

[72] Happel, J. and Brenner, H. Low Reynolds number hydrodynamics. Martinus Nijhoff Publishers, Dordrecht, (1983).

[73] Spurk, J. H. and Aksel, N. Fluid Mechanics. Springer, Berlin, 2nd edition, (2008).

[74] Aksel, N. and Schmidtchen, M. Analysis of the overall accuracy in ldv measuremet of film flow in an inclined channel. Meas. Sci. Technol. **7**, 1140–1147 (1996).

[75] Schmid, P. J. and Henningson, D. S. Stability and Transition in Shear Flows, volume 142. Springer, New York, (2001).

[76] Drazin, P. G. Introduction to Hydrodynamic Stability. Cambridge University Press, Cambridge, (2002).

[77] Taneda, S. Experimental investigation of the wakes behind cylinders and plates at low Reynolds numbers. Journal of the physical society of Japan **11**(3), 302–307 (1956).

[78] Bohr, T., Ellegaard, C., Hansen, A. E., and Haaning, A. Hydraulic jumps, flow separation and wave breaking: An experimental study. Physica B: Condensed Matter **228**(1-2), 1–10 (1996).

[79] Bohr, T., Putkaradze, V., and Watanabe, S. Averaging theory for the structure of hydraulic jumps and separation in laminar free-surface flows. Physical Review Letters **79**(6), 1038–1041 (1997).

[80] Bush, J. W. M. and Aristoff, J. M. The influence of surface tension on the circular hydraulic jump. Journal of Fluid Mechanics **489**, 229–238 (2003).

[81] Lugt, H. J. Vortex flow in nature and technology. G. Braun, New York, (1983).

[82] Dyke, M. V. An Album of Fluid Motion. The Parabolic Press, Stanford, (1997).

[83] Matlab® 2010b The Mathworks, Inc.

[84] OpenFOAM 1.6. The open source CFD Toolbox, OpenCDFD Ltd. Reading, Berkhire, UK, (2009). See http://www.openfoam.com for more information.

[85] Nyquist, H. Certain topics in telegraph transmission theory. Trans. IEEE **47**, 617–644 (1928).

[86] Schubauer, G. B. and Skramstad, H. K. Laminar-boundary-layer oscillations and transition on a flat plate. J. of the Aeronautical Sciences **14**, 69–78 (1947).

[87] Schlichting, H. and Gersten, K. Boundary-layer theory. Springer, Berlin, 8th edition, (2003).

Danksagung

Mein Dank gilt allen Mitarbeitern des Lehrstuhls für Technische Mechanik und Stömungsmechanik der Universität Bayreuth die mich bei der Erstellung dieser Arbeit tatkräftig unterstüzt haben. Hier möchte ich mich insbesondere beim Lehrstuhlinhaber Herrn Prof. Dr. Nuri Aksel bedanken, der meine Arbeit über den gesamten Zeitraum sehr geradlinig betreut hat, mir stets hilfsbereit zur Seite stand und sich immer Zeit für Diskussionen über offene Fragestellungen nahm. Regelmässig räumte er mir auch die Möglichkeit ein, an diversen internationalen Fachtagungen teilzunehmen bei denen ich sehr interessante Kontakte zu Forschungskollegen aus der ganzen Welt aufbauen und vertiefen konnte.

Außerdem hervorzuheben sind meine Kollegen Dr. André Haas, Dr. Christian Heining und Dr. Riccardo Puraglesi die mir durch zahlreiche Fachdiskussionen häufig helfen konnten neue Ideen und Lösungsansätze zu erarbeiten. Großer Dank gilt insbesondere auch Carola Lepski und Marion Märkl, da durch ihre unterstützende Laborarbeit ein zügiges und effektives wissenschaftliches Arbeiten überhaupt erst ermöglicht wurde. Weiterhin möchte ich Dr. Lutz Heymann und Katja Helmrich für ihre bereitwillige Hilfe nicht nur bei organisatorischen Fragestellungen einen großen Dank aussprechen. Selbstverständlich haben darüber hinaus auch alle anderen Mitarbeiter des Lehrstuhls zu einer Arbeitsatmosphäre beigetragen, die ich immer als sehr angenehm und kollegial empfand.

Nicht zuletzt möchte ich mich bei meinen Eltern bedanken, die mich während meiner gesamten Ausbildung uneingeschränkt unterstützt haben und mir so jeden nur erdenklichen Freiraum geschaffen haben um mein Studium und das Verfassen dieser Arbeit erfolgreich abzuschließen.

i want morebooks!

Buy your books fast and straightforward online - at one of the world's fastest growing online book stores! Environmentally sound due to Print-on-Demand technologies.

Buy your books online at
www.get-morebooks.com

Kaufen Sie Ihre Bücher schnell und unkompliziert online – auf einer der am schnellsten wachsenden Buchhandelsplattformen weltweit!
Dank Print-On-Demand umwelt- und ressourcenschonend produziert.

Bücher schneller online kaufen
www.morebooks.de

OmniScriptum Marketing DEU GmbH
Heinrich-Böcking-Str. 6-8
D - 66121 Saarbrücken
Telefax: +49 681 93 81 567-9

info@omniscriptum.de
www.omniscriptum.de

Printed by Books on Demand GmbH, Norderstedt / Germany